Wave Scattering
by Small Bodies
**Creating Materials with a Desired Refraction
Coefficient and Other Applications**

T0320700

Wave Scattering
by Small Bodies
Creating Materials with a Desired Refraction Coefficient and Other Applications

Alexander G Ramm

Kansas State University, USA

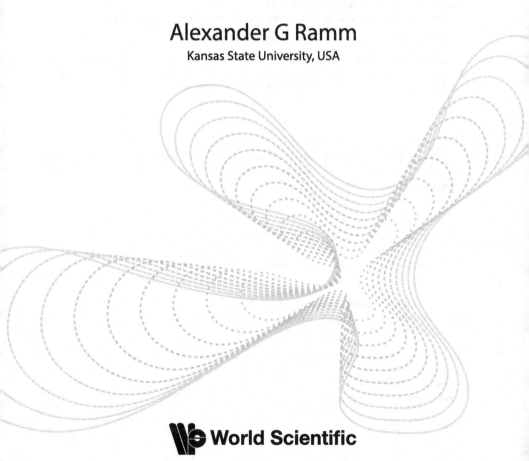

World Scientific

NEW JERSEY · LONDON · SINGAPORE · BEIJING · SHANGHAI · HONG KONG · TAIPEI · CHENNAI · TOKYO

Published by

World Scientific Publishing Co. Pte. Ltd.

5 Toh Tuck Link, Singapore 596224

USA office: 27 Warren Street, Suite 401-402, Hackensack, NJ 07601

UK office: 57 Shelton Street, Covent Garden, London WC2H 9HE

Library of Congress Control Number: 2023023147

British Library Cataloguing-in-Publication Data
A catalogue record for this book is available from the British Library.

ISBN 978-981-127-648-4 (hardcover)
ISBN 978-981-127-649-1 (ebook for institutions)
ISBN 978-981-127-650-7 (ebook for individuals)

For any available supplementary material, please visit
https://www.worldscientific.com/worldscibooks/10.1142/13415#t=suppl

Desk Editors: Soundararajan Raghuraman/Nijia Liu

Typeset by Stallion Press
Email: enquiries@stallionpress.com

Printed in Singapore

Foreword

In this monograph, the many-body wave scattering problem is solved asymptotically exactly under the assumption $a \ll d \ll \lambda$, where a is the characteristic size of a particle, d is the minimal distance between neighboring particles, and λ is the wavelength in a bounded domain in which the particles are embedded. Multiple scattering is essential in our theory.

This theory allows the author to give a recipe for creating materials with a desired refraction coefficient. This has many practical applications. In particular, one can create materials with a desired wave-focusing property.

Heat transfer theory is developed in the medium in which many small bodies are embedded.

Quantum-mechanical scattering by many potentials with small supports is studied.

Acoustic and electromagnetic wave scattering is considered under various boundary conditions.

In this edition, references to the author's papers, published after year 2013, are given. New Appendices are also added.

Alexander G. Ramm

Preface

The author wishes to let the reader know what this book is about and what practical conclusions engineers can find in this book.

In this book the author presents systematically his theory of scalar wave scattering as well as electromagnetic (EM) wave scattering by one and many small bodies of arbitrary shapes. If the characteristic size of a body is a, then smallness of the body means that $ka \ll 1$, where k is the wave number, $k = \frac{2\pi}{\lambda}$, and λ is the wavelength. In a homogeneous medium $\lambda = vT$, v is the wave velocity and T is the period of the wave, $\omega = \frac{2\pi}{T}$ is its frequency and $k = \frac{2\pi}{\lambda} = \frac{\omega}{v}$. We assume that $a \ll d \ll \lambda$.

The boundary conditions on the boundary S of a body D, which we impose, include the impedance boundary condition, the Dirichlet, the Neumann, and the transmission (interface) boundary conditions for scalar wave scattering and the impedance boundary condition for EM wave scattering.

In all cases, we give explicitly analytical formulas for the field scattered by one small body of an *arbitrary shape*. These results are new.

The theory of wave scattering by small bodies was originated by Rayleigh (1871), who understood that the main term of the scattered field under his assumptions is given by the dipole radiation. Later it was found that the magnetic dipole radiation for perfectly conducting bodies is of the same order of magnitude as the electric dipole radiation. Rayleigh and his followers did not give analytic formulas for calculating the induced dipole moment on a small body of an arbitrary shape. This was done nearly 100 years later in Ramm (1969a, 1969b, 1970, 1971a, 1971b, 1971c) and presented in the books Ramm (1980b, 1982, 2005b). The new results on wave scattering by small bodies of arbitrary shapes, presented in this book, are based on the papers Ramm (2007a, 2007b, 2007c, 2007d, 2007e, 2007f, 2007g, 2007i, 2007j, 2008a, 2008b, 2008d, 2009a, 2009b, 2009e, 2010a, 2010b, 2010c, 2011a, 2011b, 2013a, 2013b, 2013c, 2013d, 2013f). These results include an analytical and numerical solution of wave scattering by many small bodies and a derivation of the integral equation for the effective field in the medium in which many small bodies (particles) are embedded. The important point of our theory is the reduction of the many-body scattering problem to finding some numbers rather than the boundary functions. This simplifies the problem drastically and allows one to solve it. On the other hand, this reduction is asymptotically exact as $a \to 0$.

Our physical assumptions include the case when the distance d between neighboring particles is much smaller than the wavelength, $d \ll \lambda$, although $a \ll d$. These assumptions imply that the "multiple scattering effects" are crucial. By these effects we understand the influence of the field scattered by all particles, except one, on this particle. It is not possible the Born's approximation when these effects are crucial.

Although the theory of wave scattering by one and many small bodies is of great interest by itself, we illustrate its possible applications by showing how one can create a material with a desired refraction coefficient by embedding in a given material many small impedance particles. A clear recipe is formulated for doing this. Practical implementation of this recipe is discussed.

Another problem of interest is creating materials which have a desired wave-focusing property. This means that their refraction coefficient is chosen so that the corresponding radiation pattern approximates well a desired pattern, that is, a desired function on a unit sphere.

So far we have discussed mostly the scattering of scalar waves. Similar results are obtained for electromagnetic (EM) wave scattering. We derive analytical formulas for the EM field scattered by one impedance (or perfectly conducting) small body of an arbitrary shape, solve many-body EM wave scattering problems when the scatterers are small impedance bodies of an arbitrary shape, derive the equation for the effective field in the medium in which very many such bodies are embedded, and calculate the refraction coefficient in this medium.

No such results were obtained earlier, to our knowledge. These formulas are used for developing a numerical method for solving many-body scattering problems in the case of small impedance bodies of an arbitrary shape. An equation for the effective field is studied. This equation leads to new physical effects in the new medium, created by embedding many small particles. These effects include change of the refraction coefficient and of magnetic permeability. One may use these results in practice in order to change the refraction coefficient in the desired direction.

The author also discusses some physical problems of interest. For example, wave scattering by many nanowires (thin cylinders) is discussed. Heat propagation in the medium, in which many small bodies are embedded, is investigated. Theory of quantum-mechanical scattering by many potentials with small supports is developed.

Collocation method is developed for solving the equation for the effective field. Its convergence is proved and the rate of convergence is given. This collocation method is used for a justification of a homogenization-type theory. There are many books and papers dealing with homogenization theory. Quite often it is assumed in these books that periodicity condition with a parameter is satisfied, that the spectrum of the related problem is discrete, and that the corresponding operator is self-adjoint. None of these assumptions are used in our theory. We develop a version of the theory that fits our assumptions.

Several inverse problems are studied: finding the position and size of a small body from the scattering data for two scatterers, one of which is large and known; finding small subsurface inhomogeneities from the scattering data measured on the surface; inverse radiomeasurements problem: finding the EM field distribution in an aperture of a mirror antenna in radio-wave diapason from the measurements of the field scattered by a small probe.

We also derive some results in potential theory that are used in this book. Among these results some are new. For example, necessary and sufficient conditions are given for a double-layer potential to be representable as a single-layer potential, and vice versa; asymptotic of the solution to the Helmholtz equation satisfying the impedance boundary condition is given as the impedance tends to infinity; a variational principle is derived for positive quadratic forms, and some classical variational principles for electric capacitance are obtained from this abstract principle.

This book is written for a wide audience which includes mathematicians, specialists in numerical analysis, engineers and physicists. The author hopes that his theory will be implemented practically, in materials science and other areas. The reader can find some results on such a numerical implementation in Andriychuk and Ramm (2010), Andriychuk and Ramm (2011), Ramm and Andriychuk (2014b).

The book is essentially self-contained: the results necessary for understanding the details of the derivations are provided with proofs.

About the Author

 **Alexander G. Ramm** is Professor Emeritus of Mathematics at Kansas State University, USA, with broad interests in analysis, scattering theory, inverse problems, theoretical physics, engineering, signal estimation, tomography, theoretical numerical analysis, and applied mathematics.

He is an author of 718 research papers and 20 research monographs. He is also an editor of 3 books. He has lectured at many universities throughout the world, given more than 150 invited and plenary talks at various conferences, and supervised 11 PhD students. He was Fulbright Research Professor in Israel and Ukraine; distinguished visiting professor in Mexico and Egypt; Mercator Professor in Germany; Research Professor in France; and an invited plenary speaker at the 7-th PACOM. He also won the Khwarizmi international award in 2004, among others.

A G Ramm was the first to prove the uniqueness of the solution to inverse scattering problems with fixed-energy scattering data; the first to prove the uniqueness of the solution to inverse scattering problems with non-over-determined scattering data, and the first to study inverse scattering problems with under-determined scattering data. He studied inverse scattering problems for potential scattering and for scattering by obstacles. He solved many specific inverse problems and developed new methods and ideas in the area of inverse scattering problems. He introduced the notion of Property C for a pair of differential operators and applied Property C for one-dimensional and multi-dimensional inverse scattering problems.

A G Ramm solved the many-body wave scattering problem where the bodies are small particles of arbitrary shapes. He used this theory to give a recipe for creating materials with a desired refraction coefficient and materials with a desired wave-focusing property. These results attracted attention of the scientists working in nanotechnology.

A G Ramm gave formulas for the scattering amplitude for scalar and electromagnetic waves by small bodies of arbitrary shapes and analytical formulas for the polarizability tensors for these bodies. He gave a solution to the Pompeiu problem, proved Schiffer's conjecture, and gave many results about symmetry problems for PDE, including first symmetry results in harmonic analysis.

A G Ramm has developed the Dynamical Systems Method (DSM) for solving linear and non-linear operator equations, especially ill-posed. These results were used numerically and demonstrated the practical efficiency of the DSM.

A G Ramm also developed the random fields estimation theory for a wide class of random fields. He has developed a theory of convolution equations with hyper-singular integrals and solved integral equations with hyper-singular kernels analytically. He applied these results to the study of the Navier–Stokes problem (NSP). As a result, he solved the millennium problem concerning the Navier–Stokes equations. He formulated and proved the NSP paradox, which shows the contradictory nature of the NSP and the non-existence of its solution on the interval $t \; \varepsilon \; (0, \infty)$ for the initial data $v_0(x) \not\equiv 0$ and $f(x,t) = 0$.

A G Ramm has introduced a wide class of domains with non-compact boundaries. He studied the spectral properties of the Schrödinger operators in this class of domains and gave sufficient conditions for the absence of eigenvalues on the continuous spectrum of these operators.

Further, he developed the theory of local, pseudolocal, and geometrical tomography. He has proved a variety of the results concerning singularities of the Radon transform and developed multidimensional algorithms for finding discontinuities of signals from noisy discrete data.

Contents

Introduction

What is this book about? Why was it written?

This book is about wave scattering by one and many small scatters
of an arbitrary shape. The interest in this question goes back to
Rayleigh (1871), who understood that the main term in the scattered
field will be the dipole radiation. Under his assumptions the smallness
of the body means $ka \ll 1$, where $k = \frac{2\pi}{\lambda}$ is the wave number, a is
the characteristic size of the body, and the order of magnitude of
the scattering amplitude is $O(a^3)$ at a large distance from the small
body. Multiple scattering is essential in our theory. Rayleigh did not
give analytical formulas for calculating the field scattered by a small
body of an arbitrary shape. Such formulas were obtained for balls
(of an arbitrary shape) in 1908 by Mie. He did not assume that the
scatterer, the ball, is small. However, his method is based on the
separation of variables and cannot be used for bodies of an arbitrary
shape.

For small bodies of an arbitrary shape, analytical formulas allowing one to calculate the scattered field with any desired accuracy
were obtained by the author Ramm (1969b, 1970) about 100 years
after Rayleigh's work of 1871.

There is a very large literature on wave scattering by small bodies.
The reason is simple: this problem is of interest in many applications.
Let us mention light scattering by atmosphere dust, by cosmic dust,
by colloidal particles in water, finding the location and size of the
small subsurface inhomogeneities from the observation of the field,
scattered by these inhomogeneities and measured on the surface.

In this book, much attention is given to creating materials with a desired refraction coefficient by embedding many small particles into a given material so that these particles have prescribed boundary impedances and are distributed in the given material with a pre-scribed density.

It turns out that a wide range of the desired refraction coefficients can be obtained in this way. A recipe for creating a desired refraction coefficient in a given material is formulated. Namely, if an arbitrary fixed bounded domain Ω is given, the refraction coefficient of the material in this domain is known to be $n_0(x)$, outside of Ω $n_0 = $ const, and if one wishes to create in Ω *a desired refraction coefficient* $n(x)$, then the author gives a recipe for doing this. He tells how many small particles of a characteristic size a should be distributed in Ω, and with what density (number of particles per unit volume) and with what boundary impedances these small particles should be distributed in order that the resulting medium will have the refraction coefficient that differs from the desired $n(x)$ as little as one wishes.

In particular, one can create materials with negative refraction coefficient. This is of interest in connection with metamaterials.

A refraction coefficient can be created so that the resulting mate-rial will have wave-focusing property.

This means that the plane wave, scattered by the body, will have the radiation pattern which approximates well a desired function on the unit sphere. For example, the scattered field will mostly be scattered in a desired solid angle.

This book was written so that the novel methods for solving one- and many-body wave scattering problems can be available for wide audiences including mathematicians, engineers, physicists, specialists in computational mathematics, and materials science, and graduate students in these areas. The material presented is interdisciplinary, and the author tries to present this material in a self-contained way and so that it will be understandable to a reader from the aforemen-tioned broad audience.

Chapter 1

Scalar Wave Scattering by One Small Body of an Arbitrary Shape

1.1 Impedance bodies

Consider the following wave scattering problem:

$$(\nabla^2 + k^2)u = 0 \quad \text{in } D' := \mathbb{R}^3 \setminus D, \tag{1.1.1}$$

$$u_N^- = \zeta u \quad \text{on } S, \tag{1.1.2}$$

$$u = u_0 + v, \tag{1.1.3}$$

$$\frac{\partial v}{\partial |x|} - ikv = o\left(\frac{1}{|x|}\right), \quad |x| \to \infty. \tag{1.1.4}$$

Here, $k > 0$ is a wave number, $k = \text{const}$, D is a bounded domain with a smooth boundary S, N is the unit normal to S pointing out of D, u_N^- is the limiting value of the normal derivative of u on S from D', ζ is a constant which we call the boundary impedance, $u_0 = e^{ik\alpha x}$ is the incident field, the plane wave, $\alpha \in S^2$, is a unit vector, S^2 is the unit sphere, v is the scattered field, condition (1.1.4) is the radiation condition at infinity. It is assumed to be satisfied uniformly with respect to the unit vector $x^0 = \frac{x}{|x|}$, that is, with respect to the direction along which x tends to infinity. Throughout this book, smoothness of S means that the equation of the surface S in local coordinates is a $C^{1,\lambda}$ function. The local coordinates are defined as follows. Fix a point $s \in S$ and let it be the origin of the coordinate system, the z-axis of which is directed along the normal N_s to S

1

at the point s, and the plane x, y is tangent to S at this point. We often write x_1, x_2, x_3 in place of x, y, z, and then $x = (x_1, x_2, x_3)$. Let $x_3 = f(x_1, x_2)$ be the equation of S in this coordinate system. Then, by construction, $f(0,0) = 0$, $\frac{\partial f(0,0)}{\partial x_j} = 0$, $j = 1, 2$. The assumption $S \in C^{1,\lambda}, 0 < \lambda \le 1$, means that $|\nabla f(x_1, x_2) - \nabla f(x_1', x_2')| \le C[(x_1 - x_1')^2 + (x_2 - x_2')^2]^{\lambda/2}$, where C is a constant that does not depend on s, x, and x', and $\nabla = e_1 \frac{\partial}{\partial x_1} + e_2 \frac{\partial}{\partial x_2}$, $\{e_j\}_{j=1}^3$ is the Cartesian basis of $\mathbb{R}^3$.

The scattering problems (1.1.1)–(1.1.4) can be considered, for example, as the scattering of an acoustic wave. In this case, u has the physical meaning of the pressure or acoustic potential. The Dirichlet boundary condition $u|_S = 0$ describes an acoustically soft body (zero pressure on the boundary) and the Neumann boundary condition $u_N|_S = 0$ describes an acoustically hard body (the normal component of the velocity ∇u vanishes on S). The impedance boundary condition describes a linear relation between the pressure and the normal component of the velocity on S.

The first task is to prove that the scattering problem has a solution and this solution is unique. In this case, we say that the scattering problem has a unique solution. If we want to state that there is at most one solution, but the existence of the solution is not asserted, then we say there exists at most one solution.

Theorem 1.1.1. *Assume that Im$\zeta \le 0$. Then problems* (1.1.1)– (1.1.4) *have at most one solution.*

Proof. Since the problem is linear, it is sufficient to prove that the corresponding homogeneous problem, that is, the problem with $u_0 = 0$, has only the trivial solution $u = 0$. To prove this, multiply equation (1.1.1) by $\bar{u}$, the complex conjugate of u, subtract the complex conjugate equation (1.1.1) multiplied by u and integrate over the region $D' \cap B_R := D'_R$, where $R > 0$ is a large number that we take to infinity and $B_R = \{x : |x| \le R\}$ is a ball of radius R centered at the origin. The origin we take inside D arbitrarily. The result is

$$0 = \int_{D'_R} [\bar{u}(\nabla^2 + k^2)u - u(\nabla^2 + k^2)\bar{u}]dx = \int_{S_R} (\bar{u}u_N - u\bar{u}_N)ds$$

$$- \int_S (\bar{u}u_N - u\bar{u}_N)ds \tag{1.1.5}$$

Here, Green's formula was used.

From the radiation condition (1.1.4), one gets:

$$\int_{S_R} (\bar{u}u_N - u\bar{u}_N)ds = 2ik \int_{S_R} |u|^2 ds + o(1), \quad R \to \infty. \qquad (1.1.6)$$

From the impedance boundary condition (1.1.2), one gets:

$$-\int_S (\bar{u}u_N - u\bar{u}_N)ds = \int_S (-\zeta|u|^2 + \bar{\zeta}|u|^2)ds = -2i\mathrm{Im}\zeta \int_S |u|^2 ds. \qquad (1.1.7)$$

From (1.1.5)–(1.1.7), it follows that

$$\lim_{R\to\infty} \left(k \int_{S_R} |u|^2 ds - \mathrm{Im}\zeta \int_S |u|^2 ds \right) = 0. \qquad (1.1.8)$$

If $\mathrm{Im}\zeta \leq 0$, then relation (1.1.8) implies

$$\lim_{R\to\infty} \int_{S_R} |u|^2 ds = 0. \qquad (1.1.9)$$

This and the radiation condition (1.1.4) imply that $u = 0$ outside any ball $B_R \supset D$. This claim is proved in Ramm (1986), p. 25. It is known as Rellich's lemma. Theorem 1.1.1 is now a consequence of the unique continuation principle for a solution to the homogeneous Helmholtz equation: if such a solution vanishes on an open subset in the domain D' where it solves the homogeneous Helmholtz equation, then this solution vanishes everywhere in D'.

Theorem 1.1.1 is proved. $\square$

Remark 1.1.1. Physically the impedance ζ can be any constant satisfying the condition $\mathrm{Im}\zeta \leq 0$, which guarantees the uniqueness of the solution to the scattering problems (1.1.1)–(1.1.4).

Let us now prove the existence of the solution to the scattering problems (1.1.1)–(1.1.4).

Theorem 1.1.2. *Problems (1.1.1)–(1.1.4) have a (unique) solution. This solution can be found of the form*

$$u(x) = u_0(x) + \int_S g(x,t)\sigma(t)dt, \qquad (1.1.10)$$

$$g(x,y) := \frac{e^{ik|x-y|}}{4\pi|x-y|}. \tag{1.1.11}$$

The function $\sigma(t)$ in (1.1.10) is uniquely determined.

Proof. For any σ function (1.1.10) solves equation (1.1.1) and satisfies conditions (1.1.3)–(1.1.4). Therefore, it solves problems (1.1.1)–(1.1.4) if and only if σ can be found so that condition (1.1.2) is satisfied, that is,

$$u_{0N} - \zeta u_0 + \frac{A\sigma - \sigma}{2} - \zeta T\sigma = 0, \tag{1.1.12}$$

where

$$A\sigma := 2 \int_S \frac{\partial g(s,t)}{\partial N_s} \sigma(t) dt, \quad T\sigma := \int_S g(s,t)\sigma(t) dt. \tag{1.1.13}$$

Formula

$$\frac{\partial}{\partial N_s} \int_S g(x,t)\sigma(t) dt \bigg|_{x \in D', x \to s} = \frac{A\sigma - \sigma}{2} \tag{1.1.14}$$

is proved in Section 11.1, see also Günter (1967). Equation (1.1.12) is of Fredholm type because operators A and T are compact in $C(S)$, see Appendix A. Therefore, the existence of the solution to equation (1.1.12) follows from the Fredholm alternative (see Appendix A) if one proves that the homogeneous equation (1.1.12) has only the trivial solution. The homogeneous equation (1.1.12) is of the form

$$\frac{A\sigma - \sigma}{2} - \zeta T\sigma = 0, \tag{1.1.15}$$

which is

$$u_N^- - \zeta u = 0 \quad \text{on } S. \tag{1.1.16}$$

By Theorem 1.1.1, the corresponding function $u(x) = 0$ in D' because it solves equation (1.1.1) in D' and satisfies the radiation condition (1.1.4) and the boundary condition (1.1.2).

If $u = 0$ in D', then $u = 0$ on S because potentials of a single layer are continuous in $\mathbb{R}^3$, see Section 11.1. Thus, u solves equation (1.1.1) in D and satisfies condition (1.1.2). This implies that $u = 0$ in D.

Indeed, multiply equation (1.1.1) in D by $\bar{u}$, use Green's formula and the boundary condition (1.1.2), and get

$$2i\mathrm{Im}\zeta \int_S |u|^2 ds = 0. \qquad (1.1.17)$$

If $\mathrm{Im}\zeta \neq 0$, then $u = 0$ on S, so $u_N = 0$ on S and $u = 0$ in D by the uniqueness of the solution to the Cauchy problem for the Helmholtz equation. If $u = 0$ in $D \cup D'$, then $\sigma = u_N^+ - u_N^- = 0$, and Theorem 1.1.2 is proved. If $\mathrm{Im}\zeta = 0$, then one deals with the real number ζ in (1.1.2).

If, as we assume, the body D is sufficiently small, then k^2 is not an eigenvalue of the Laplacian in D also in the case $\mathrm{Im}\zeta = 0$. So, in this case also $u = 0$ in $D \cup D'$ and $\sigma = 0$.

Theorem 1.1.2 is proved. □

The existence and uniqueness of the solution to the scattering problem does not depend on the size of the body D. This size we measure by the number $a := \frac{1}{2}\mathrm{diam}D$. We assume that the body is small in the sense $ka \ll 1$. This assumption allows us to derive analytic, closed form formulas for the field, scattered by D, at the distances $d \gg a$, in the "far zone".

Our next task is to derive an analytical formula for the scattered field at the distance $|x| \gg a$. To do this, take into account the following formula:

$$\frac{e^{ik|x-t|}}{4\pi|x-t|} = \frac{e^{ik|x|-ikx^0 \cdot t}}{4\pi|x|}\left(1 + O\left(\frac{|t|}{|x|}\right)\right), \quad |t| \leq a. \qquad (1.1.18)$$

Here, $x^0 = \frac{x}{|x|}$. Formula (1.1.18) can be established easily. One has

$$|x-t| = |x|\sqrt{1 - \frac{2x^0 \cdot t}{|x|} + \frac{|t|^2}{|x|^2}} = |x|\left(1 + O\left(\frac{|t|}{|x|}\right)\right) \qquad (1.1.19)$$

and

$$e^{ik|x-t|} = e^{ik|x|-ikx^0 \cdot t + O\left(\frac{|t|^2}{|x|}\right)} = e^{ik|x|-ikx^0 \cdot t}\left(1 + O\left(\frac{|t|^2}{|x|}\right)\right). \qquad (1.1.20)$$

If $|t| \leq a$ and a is small, then $O\left(\frac{|t|^2}{|x|}\right) \leq O\left(\frac{|t|}{|x|}\right)$, so (1.1.18) holds. In what follows we denote $x^0 = \beta$ and $|x| = r$. One has

$$\int_S g(x,t)\sigma(t)dt = \frac{e^{ikr}}{4\pi r}\int_S e^{-ik\beta \cdot t}\sigma(t)dt\left(1 + O\left(\frac{a}{r}\right)\right). \qquad (1.1.21)$$

Let $g(r) = \frac{e^{ikr}}{4\pi r}$ and

$$Q := \int_S \sigma(t)dt. \qquad (1.1.22)$$

If

$$|Q| \gg \left|\int_S t\sigma(t)dt\right|, \qquad (1.1.23)$$

then one can write

$$\int_S e^{-ik\beta \cdot t}\sigma(t)dt \sim \int_S \sigma(t)dt = Q, \qquad (1.1.24)$$

where *the sign $\sim$ stands for the asymptotic equality as $a \to 0$.*
   If (1.1.23) holds, then

$$u(x) \sim u_0(x) + g(r)Q, \quad r = |x| \gg a. \qquad (1.1.25)$$

Therefore, the solution of the scattering problem is completed if the number Q is found.
   To find Q, we use the exact integral equation (1.1.12). However, we do not solve it for σ, but find the main term of Q as $a \to 0$ asymptotically, in closed form. To do this, integrate equation (1.1.12) over S and evaluate the significance of each term as $a \to 0$.
   The first term is

$$I_1 := \int_S u_{0N}ds = \int_D \nabla^2 u_0 dx = -k^2\int_D u_0 dx = O(a^3), \qquad (1.1.26)$$

where we took into account that

$$\left|\int_D u_0 dx\right| \leq \int_D dx = |D| = O(a^3), \qquad (1.1.27)$$

and $|D| := V$ is the volume of D.

The second term is

$$I_2 := \int_S \zeta u_0 ds = O(a^2), \tag{1.1.28}$$

where

$$\int_S ds = |S| = O(a^2), \tag{1.1.29}$$

and $|S|$ is the surface area of S.

The third term is

$$I_3 := \int_S \frac{A\sigma - \sigma}{2} ds = -\frac{Q}{2} + \frac{1}{2} \int_S A\sigma ds. \tag{1.1.30}$$

We claim that

$$\int_S A\sigma ds \sim \int_S A_0 \sigma ds = -\int_S \sigma ds = -Q, \tag{1.1.31}$$

where

$$A_0\sigma := 2 \int_S \frac{\partial g_0(s,t)}{\partial N_s} \sigma(t) dt, \quad g_0(x,t) := \frac{1}{4\pi|x-t|}. \tag{1.1.32}$$

Indeed, one has

$$
\begin{aligned}
A\sigma &= 2 \int_D \frac{\partial}{\partial N_s} \frac{e^{ik|s-t|}}{4\pi|s-t|} \sigma(t) dt \\
&= 2 \int_S \frac{\partial}{\partial N_s} \left(\frac{1 + ik|s-t| + O(k^2|s-t|^2)}{4\pi|s-t|} \right) \sigma(t) dt \\
&= A_0\sigma + \int_S O(|s-t|)\sigma(t) dt \\
&\sim A_0\sigma \quad \text{as } a \to 0.
\end{aligned}
\tag{1.1.33}
$$

Let us check that

$$\int_S A_0 \sigma ds = -\int_S \sigma dt. \tag{1.1.34}$$

This is an exact relation. One has

$$2 \int_S ds \int_S \frac{\partial}{\partial N_S} \frac{1}{4\pi |s - t|} \sigma(t) dt = \int_S dt \sigma(t) 2 \int_S \frac{\partial}{\partial N_S} \frac{1}{4\pi |s - t|} ds$$

$$= - \int_S \sigma(t) dt. \qquad (1.1.35)$$

Here we have used the known formula (see Section 11.1):

$$2 \int_S \frac{\partial}{\partial N_S} \frac{1}{4\pi |s - t|} ds = -1, \quad t \in S. \qquad (1.1.36)$$

Proof of this formula is given in Section 11.1.
Thus,

$$I_3 \sim -Q, \quad \text{as } a \to 0. \qquad (1.1.37)$$

Finally,

$$I_4 := \varsigma \int_S T\sigma ds = \int_S dt \sigma(t) \varsigma \int_S g(s, t) ds. \qquad (1.1.38)$$

One has

$$\int_S g(s, t) ds = O(a), \quad a \to 0. \qquad (1.1.39)$$

Let us assume that

$$\lim_{a \to 0} \varsigma a = 0. \qquad (1.1.40)$$

Assuming (1.1.40), we allow ς to depend on a in such a way that (1.1.40) holds. If ς is a constant independent of a, then (1.1.40) is obviously valid. If (1.1.40) holds, then

$$I_4 = o(Q), \quad a \to 0, \qquad (1.1.41)$$

provided that $Q \neq 0$. We will prove that $Q \neq 0$ under our assumptions. Keeping the terms I_3 and I_2 and neglecting the terms I_1 and I_4

of higher order of smallness as $a \to 0$, one gets

$$Q \sim -\zeta |S| u_0(x_1), \quad x_1 \in D. \tag{1.1.42}$$

If $u_0(x) = e^{ik\alpha \cdot x}$ and $x_1 \in D$, then $u_0(x_1) = 1 + O(ka)$, since $ka \ll 1$.

Formula (1.1.42) is our *basic result*. From (1.1.42) and (1.1.25) one gets an analytical formula for the solution of the wave scattering problems (1.1.1)–(1.1.4) in the case of a small body of an arbitrary shape:

$$u(x) \sim u_0(x) - g(r)\zeta |S|, \quad |x| \gg a, \tag{1.1.43}$$

where $g(r) = \frac{e^{ikr}}{4\pi r}$, $r = |x|$, and $u_0(x_1) \sim 1$ if $ka \ll 1$.

It is assumed in (1.1.43) that the origin is inside D. If the origin is elsewhere, and $x_1 \in D$, then formula (1.1.43) takes the form

$$u(x) \sim u_0(x) - g(|x - x_1|)\zeta |S|, \quad |x - x_1| \gg a. \tag{1.1.44}$$

Let us define the scattering amplitude $A(\beta, \alpha, k)$ by the formula

$$v \sim \frac{e^{ikr}}{r} A(\beta, \alpha, k) + o\left(\frac{1}{r}\right), \quad r = |x| \to \infty, \quad \frac{x}{r} := \beta. \tag{1.1.45}$$

Comparing (1.1.43) and (1.1.45), one obtains

$$A(\beta, \alpha, k) = -\frac{\zeta |S|}{4\pi} = \frac{Q}{4\pi}, \tag{1.1.46}$$

where it is assumed that $x_1 = 0$, so that $u_0(x_1) = 1$.

The physical conclusion is: the scattering is isotropic, that is, it does not depend on α and β, and in absolute value the scattering amplitude is $O(\zeta |S|) = O(\zeta a^2)$.

Let us summarize the results in a theorem.

Theorem 1.1.3. *Assume that $ka \ll 1$, $Im\zeta \leq 0$, and (1.1.40) holds. Then the scattering problems (1.1.1)–(1.1.4) has a unique solution. This solution can be calculated by formula (1.1.43) if the origin is inside D, and by formula (1.1.44) if the origin is elsewhere and $x_1 \in D$.*

The scattering is isotropic and the scattering scattering amplitude is $O(\zeta a^2)$.

1.2 Acoustically soft bodies (the Dirichlet boundary condition)

In this section, problems (1.1.1), (1.1.3), and (1.1.4) are studied, but the impedance boundary condition (1.1.2) is replaced by the Dirichlet condition

$$u = 0 \quad \text{on } S. \tag{1.2.1}$$

Physically this condition in acoustics means that the body is acoustically soft, that is, the pressure on S is vanishing. Mathematically condition (1.2.1) is the limiting case of condition (1.1.2) when $\zeta \to \infty$. This is proved in Section 11.3. The case $\zeta \to 0$ yields the Neumann condition $u_N = 0$ on S, corresponding to an acoustically hard body.

The scattering problem with the Dirichlet condition has a unique solution, as we will prove, and the analogs of Theorems 1.1.1–1.1.3 will be established.

Theorem 1.2.1. *Problems* (1.1.1), (1.2.1), *and* (1.1.3)–(1.1.4) *have at most one solution.*

Proof.   It is sufficient to prove that the corresponding homogeneous problem, that is, the problem with $u_0 = 0$, has only the trivial solution. The proof is essentially the same as the proof of Theorem 1.1.1.

One obtains an analog of relation (1.1.8) with the integral over S vanishing because $u = 0$ on S. Therefore, one concludes that relation (1.1.9) holds.

The rest of the proof goes as in the proof of Theorem 1.1.1.

Theorem 1.2.1 is proved.   □

Let us prove the existence of the solution to problems (1.1.1), (1.2.1), (1.1.3), and (1.1.4).

Theorem 1.2.2. *The above problem has a solution.*

Proof.   Let us look for the solution of the form (1.1.10). As in the proof of Theorem 1.1.2 it is sufficient to prove that the equation that one obtains from the boundary condition (1.2.1) is solvable.

This equation is

$$\int_S g(s,t)\sigma(t)dt = -u_0(s), \quad s \in S. \tag{1.2.2}$$

It is proved in Section 11.2 that equation (1.2.2) is (uniquely) solvable provided that k^2 is not an eigenvalue of the Dirichlet Laplacian in D. This condition is satisfied if D is sufficiently small.

Theorem 1.2.2 is proved. □

Let us prove an analog of Theorem 1.1.3.

As in Section 1.1, one gets formula (1.1.25) where Q is defined in (1.1.22). To find an analytical expression for Q, we rewrite equation (1.2.2) as

$$\int_S g_0(s,t)\sigma(t)dt = -1 + O(ka), \quad g_0(s,t) = \frac{1}{4\pi|s-t|}. \tag{1.2.3}$$

Neglecting the small term $ka \ll 1$, one gets

$$\int_S g_0(s,t)\sigma(t)dt = -1, \quad s \in S. \tag{1.2.4}$$

This is an equation for the surface charge distribution σ that generates the constant potential $\mathcal{U}$ on S, $\mathcal{U} = -1$. With this interpretation the body D is a perfect conductor and the quantity $Q = \int_S \sigma(t)dt$ is its total charge. There is a well-known relation $C\mathcal{U} = Q$, where C is the capacitance of the perfect conductor D. Thus,

$$Q = -C, \tag{1.2.5}$$

and formula (1.1.25) takes the form

$$u(x) = u_0(x) - g(r)C, \quad r = |x| \gg a, \tag{1.2.6}$$

while formula (1.1.46) becomes

$$A(\beta,\alpha,k) = -\frac{C}{4\pi}. \tag{1.2.7}$$

These formulas solve the scattering problems (1.1.1), (1.2.1), and (1.1.3), (1.1.4).

Let us formulate the results.

Theorem 1.2.3. *If $ka \ll 1$, then the solution to the above scattering problem exists, is unique, can be calculated by formula (1.2.6), and the scattering amplitude is given by formula (1.2.7).*

Remark 1.2.1. This result is very useful practically because the author has derived explicit analytical formulas which allow one to calculate electrical capacitance C for a conductor of an arbitrary shape, see Ramm (2005b) and Section 11.4.

For example, the zeroth approximation of C is given by the following formula:

$$C^{(0)} = \frac{4\pi|S|^2}{\int_S \int_S \frac{dsdt}{|s-t|}}, \quad C^{(0)} \leq C, \tag{1.2.8}$$

where $|S|$ is the surface area of S and the electric permittivity in D' is equal to 1.

Let us draw some physically interesting conclusions from Theorem 1.2.3.

Note that the electrical capacitance $C = O(a)$. Formula (1.2.7) shows that the scattering is isotropic and the scattering amplitude is $O(a)$, that is, it is *much larger* than in the case of the impedance boundary condition. If, for example, $\zeta = O(\frac{1}{a^\kappa}), 0 \leq \kappa < 1$, so that condition (1.1.40) is satisfied, then formula (1.1.46) yields $A(\beta, \alpha, k) = O(a^{2-\kappa})$, and $O(a^{2-\kappa}) \ll O(a)$ if $\kappa < 1$ and $a \to 0$.

1.3 Acoustically hard bodies (the Neumann boundary condition)

Consider now the scattering problems (1.1.1), (1.1.3), and (1.1.4) and replace condition (1.1.2) by the Neumann boundary condition

$$u_N = 0 \quad \text{on } S. \tag{1.3.1}$$

Our plan of the study is unchanged: we want to prove existence of a unique solution and derive analytical formulas for the solution and for the scattering amplitude.

Theorem 1.3.1. *Problems (1.1.1), (1.3.1), (1.1.3), and (1.1.4) have at most one solution.*

Proof. One can use the proof of Theorem 1.1.1 taking $\zeta = 0$ in this proof.

Theorem 1.3.1 is proved.   □

Theorem 1.3.2. *The scattering problems* (1.1.1), (1.3.1), (1.1.3), *and* (1.1.4) *has a solution.*

Proof. Let us look for the solution of the form (1.1.10). As in the proof of Theorem 1.1.2 it is sufficient to prove that equation (1.1.12) with $\zeta = 0$ has a solution. Since equation (1.1.12) with $\zeta = 0$ is of Fredholm type, the existence of its solution will be proved if one proves that equation (1.1.15) with $\zeta = 0$ has only the trivial solution. Let us prove this. The equation

$$\frac{A\sigma - \sigma}{2} = 0 \qquad (1.3.2)$$

implies that $u_N^- = 0$ on S. By Theorem 1.3.1, it follows that $u = 0$ in D'. Therefore, $u = 0$ on S and u solves the Dirichlet problem for equation (1.1.1) in D. This implies that $u = 0$ in D because k^2 cannot be a Dirichlet eigenvalue of the Laplacean if D is sufficiently small. If $u = 0$ in $D \cup D'$, then $\sigma = u_N^+ - u_N^- = 0$. By the Fredholm alternative, the existence of the solution to equation (1.1.12) with $\zeta = 0$ is proved.

Theorem 1.3.2 is proved.   □

Let us now prove an analog of Theorem 1.1.3.

Formula (1.1.24) now takes the form

$$Q_1 := \int_S e^{-ik\beta \cdot t} \sigma(t)dt \sim \int_S \sigma(t)dt - ik\beta_p \int_S t_p \sigma(t)dt, \qquad (1.3.3)$$

where *here and subsequently over the repeated indices our summation is understood.*

The novel point in a study of the scattering problem with the Neumann boundary condition is the necessity to use both terms (1.3.3) because they are of the same order as $a \to 0$, namely, they are both $O(a^3)$. This is in contrast with the problem with the Dirichlet boundary condition where $Q = O(a)$, and with the impedance boundary condition where $Q = O(\zeta a^2)$ as $a \to 0$. Moreover, we prove that the scattering in the problem with Neumann boundary condition is *anisotropic*, in sharp contrast with the cases of the Dirichlet and impedance boundary conditions.

Let us first estimate

$$Q = \int_S \sigma(t)dt.$$

We look for the solution to problems (1.1.1), (1.3.1), (1.1.3), and (1.1.4) of the form (1.1.10). The boundary condition (1.3.1) yields the integral equation for σ as follows:

$$\sigma = A\sigma + 2u_{0N} \quad \text{on } S. \tag{1.3.4}$$

Integrate this equation over S and use formula (1.1.30) to get

$$Q \sim \int_S u_{0N}ds = \int_D \nabla^2 u_0 dx = \nabla^2 u_0(x_1)|D|, \tag{1.3.5}$$

where $x_1 \in D$ is an arbitrary point inside D and we took into account that because D is small, $\nabla^2 u_0(x_1)$ does not depend on the choice of x_1. Since $D = O(a^3)$, it follows from (1.3.5) that $Q = O(a^3)$ as was mentioned earlier.

Let us estimate the last integral in (1.3.3). Introduce the matrix (tensor) β_{pq} by the formula

$$\beta_{pq} := \frac{1}{|D|} \int_S t_p \sigma_q(t)dt, \tag{1.3.6}$$

where σ_q is the unique solution to the equation

$$\sigma_q = A\sigma_q - 2N_q, \quad q = 1, 2, 3, \tag{1.3.7}$$

and $N_q = N \cdot e_q$, where $\{e_q\}$ is an orthonormal Cartesian basis of $\mathbb{R}^3$. Equation (1.3.7) is of Fredholm type and it has a solution because the corresponding homogeneous equation $\sigma = A\sigma$ has only the trivial solution $\sigma = 0$. Indeed, this equation is equivalent to the relation $u_N^- = 0$, and $u = \int_S g(x,t)\sigma(t)dt$ satisfies equation (1.1.1) and the radiation condition (1.1.4). Therefore, $u = 0$ in D'. Consequently, $u = 0$ on S. Thus, u solves equation (1.1.1) in D and vanishes on S. This means that k^2 is the Dirichlet eigenvalue of the Laplacian in D, which is a contradiction because D is assumed to be sufficiently small. Therefore, $u = 0$ in $D \cup D'$, and $\sigma = u_N^+ - u_N^- = 0$. This proves the existence and uniqueness of the solution σ_q to equation (1.3.7). The function σ in equation (1.3.3) solves equation (1.3.4), where

$$u_{0N} = \frac{\partial u_0}{\partial x_q} N_q. \tag{1.3.8}$$

Recall that summation over q is understood. Thus,

$$2u_{0N} = -2N_q \left(-\frac{\partial u_0(x_1)}{\partial x_q} \right). \tag{1.3.9}$$

From (1.3.4), (1.3.9), (1.3.7), and (1.3.6), one gets

$$-ik\beta_p \int_S t_p\sigma(t)dt = ik\beta_{pq}\beta_p \frac{\partial u_0(x_1)}{\partial x_q}|D|, \tag{1.3.10}$$

where $|D|$ is the volume of D, β_{pq} is the tensor defined in (1.3.6), and β_p is the pth component of the unit vector $\beta := \frac{x-x_1}{|x-x_1|}$, $x_1 \in D$.
    Therefore,

$$Q_1 \sim |D| \left(\nabla^2 u_0(x_1) + ik\beta_{pq}\beta_p \frac{\partial u_0(x_1)}{\partial x_q} \right), \quad x_1 \in D, \tag{1.3.11}$$

where Q_1 is defined in (1.3.3).
    Consequently, an analog of formulas (1.1.44) and (1.1.45) in the case of the Neumann boundary condition takes the form

$$u(x) = u_0(x) + g(x, x_1) \left(\nabla^2 u_0(x_1) + ik\beta_{pq}\beta_p \frac{\partial u_0(x_1)}{\partial x_q} \right) |D|,$$

$$|x - x_1| \gg a, \tag{1.3.12}$$

where

$$\beta = \frac{x - x_1}{|x - x_1|}, \quad x_1 \in D, \tag{1.3.13}$$

and

$$A(\beta, \alpha, k) = \left(\nabla^2 u_0(x_1) + ik\beta_{pq}\beta_p \frac{\partial u_0(x_1)}{\partial x_q} \right) \frac{|D|}{4\pi}. \tag{1.3.14}$$

If x_1 is the origin, $x_1 = 0$, and $u_0 = e^{ik\alpha \cdot x}$, then formula (1.3.14) can be rewritten as

$$A(\beta, \alpha, k) = -\frac{k^2|D|}{4\pi} (1 + \beta_{pq}\beta_p\alpha_q). \tag{1.3.15}$$

One can see from this formula that $A = O(a^3)$, the scattering is anisotropic and its anisotropic part is described by the tensor β_{pq} defined by formula (1.3.6).

Let us summarize the results.

Theorem 1.3.3. *The scattering problems* (1.1.1), (1.3.1), *and* (1.1.3)–(1.1.4) *is uniquely solvable. Its solution can be calculated by formulas* (1.3.12)–(1.3.13).

The scattering amplitude $A(\beta, \alpha, k)$ *is given by formula* (1.3.14), $A(\beta, \alpha, k) = O(a^3)$, *and the scattering is anisotropic.*

1.4   The interface (transmission) boundary condition

Consider the following scattering problem:

$$(\nabla^2 + k^2)u = 0 \quad \text{in } D', \quad k^2 > 0, \tag{1.4.1}$$

$$(\nabla^2 + k_1^2)u = 0 \quad \text{in } D, \quad k_1^2 > 0, \tag{1.4.2}$$

$$u^+ = u^-, \quad \rho_1 u_N^+ = u_N^-, \quad \rho_1 > 0, \tag{1.4.3}$$

$$u = u_0 + v, \tag{1.4.4}$$

$$\frac{\partial v}{\partial r} - ikv = o\left(\frac{1}{r}\right), \quad r = |x| \to \infty, \tag{1.4.5}$$

where u_0 is the incident field,

$$(\nabla^2 + k^2)u_0 = 0 \quad \text{in } \mathbb{R}^3. \tag{1.4.6}$$

In particular, the plane wave incident field $u_0 = e^{ik\alpha \cdot x}, \alpha \in S^2$, is often considered. The number ρ_1 in the transmission boundary condition (1.4.3) is assumed positive and $\rho_1 \neq 1$. The number k_1^2 in (1.4.2) is not equal to k^2.

Physically, problems (1.4.1)–(1.4.6) corresponds to the incident wave propagating through D and in this process the wave is scattered.

As earlier, the first task is to prove the existence and uniqueness of the solution to the scattering problems (1.4.1)–(1.4.6).

Theorem 1.4.1. *The above problem with* $\rho_1 = \text{const} > 0$ *has no more than one solution.*

Proof.   It is sufficient to prove that the homogeneous problem, that is, the problem with $u_0 = 0$, has only the trivial solution. To prove this, multiply equations (1.4.1) and (1.4.2) by $\bar{u}$, the bar stands for complex conjugate, integrate the second equation over D, the first equation over $D'_R := B_R \cap D'$, $B_R := \{x : |x| \leq R\}$, then use Green's formula and conditions (1.4.3) and (1.4.5) and get

$$\int_S (\bar{u}u_N^+ - u\overline{u_N^+})ds - \int_S (\bar{u}u_N^- - u\overline{u_N^-})ds + \int_{S_R} (\bar{u}u_N - u\overline{u_N})ds$$

$$= \int_S [\bar{u}u_N^+(1 - \rho_1) - u\overline{u_N^+}(1 - \rho_1)]ds + 2ik \int_{S_R} |u|^2 ds + o(1)$$

$$= 0, \tag{1.4.7}$$

where $\lim_{R \to \infty} o(1) = 0$.

One has, using Green's formula,

$$(1 - \rho_1) \int_S (\bar{u}u_N^+ - u\overline{u_N^+})ds = 0, \tag{1.4.8}$$

because u and $\bar{u}$ solve the same equation (1.4.2).

Thus, relation (1.4.7) implies

$$\lim_{R \to \infty} \int_{S_R} |u|^2 ds = 0. \tag{1.4.9}$$

From this, equation (1.4.1), and the radiation condition (1.4.5) with $u_0 = 0, u = v$, it follows that $u = 0$ in D'. Thus, $u^- = u_N^- = 0$.

Consequently, $u^+ = u_N^+ = 0$, and $u = 0$ in D.

Theorem 1.4.1 is proved. □

Remark 1.4.1. The proof remains valid with only a slight change if one assumes that $\text{Im}k^2 \geq 0$.

Theorem 1.4.2. *Problems* (1.4.1)–(1.4.6) *have a solution.*

Proof.   Let us look for a solution of the form

$$u(x) = u_0(x) + \int_S g(x, t)\sigma(t)dt + \kappa \int_D g(x, y)u(y)dy, \tag{1.4.10}$$

where

$$\kappa := k_1^2 - k^2, \quad g(x,y) = \frac{e^{ik|x-y|}}{4\pi|x-y|}, \qquad (1.4.11)$$

and σ has to be found such that conditions (1.4.3) are satisfied.

Note that equations (1.4.1), (1.4.2), (1.4.4), (1.4.5), and (1.4.6) are satisfied.

This is obvious for equations (1.4.4)–(1.4.6) and follows for equations (1.4.1)–(1.4.2) from the equation

$$(\nabla^2 + k^2)g(x,y) = -\delta(x-y). \qquad (1.4.12)$$

The first condition (1.4.3) is satisfied because both integrals in (1.4.10) are continuous functions in $\mathbb{R}^3$.

The second condition (1.4.3) is satisfied if σ and u solve the system of integral equations (1.4.10) and (1.4.13), where

$$\rho_1 \frac{A\sigma + \sigma}{2} - \frac{A\sigma - \sigma}{2} + (\rho_1 - 1)\frac{\partial}{\partial N_S}Bu + (\rho - 1)u_{0N} = 0. \quad (1.4.13)$$

Here,

$$A\sigma = 2\int_S \frac{\partial g(s,t)}{\partial N_S}\sigma(t)dt, \quad Bu = \kappa\int_D g(x,y)u(y)dy. \qquad (1.4.14)$$

Let us rewrite equation (1.4.13) as follows:

$$\sigma = \lambda A\sigma + 2\lambda B_1 u + 2\lambda u_{0N}, \qquad (1.4.15)$$

where

$$\lambda = \frac{1-\rho}{1+\rho}, \quad B_1 u := \frac{\partial(Bu)}{\partial N_S}. \qquad (1.4.16)$$

If $0 < \rho < \infty$, then $\lambda \in (-1,1)$. We assume, as earlier, that

$$ka \ll 1. \qquad (1.4.17)$$

If the pair $\{\sigma, u(x)|_{x\in D}\}$ is found from the system of integral equations (1.4.10) and (1.4.15), then the solution to the scattering problems (1.4.1)–(1.4.6) is found in $\mathbb{R}^3$ by formula (1.4.10). $\qquad \square$

Theorem 1.4.3. *The systems* (1.4.10)–(1.4.15) *have a unique solution. Thus, the scattering problems* (1.4.1)–(1.4.6) *has a unique solution.*

Proof. Equations (1.4.10)–(1.4.15) are of Fredholm type. Therefore, it is sufficient to prove that the system of homogeneous versions of these equations has only the trivial solution.

From the derivation of equations (1.4.10) and (1.4.15), it follows that the homogeneous versions of these equations imply that u solves the scattering problems (1.4.1)–(1.4.6) with $u_0 = 0$.

By Theorem 1.4.1, this problem has only the trivial solution $u = 0$. Therefore the system of equations (1.4.10) and (1.4.15) has a solution and this solution is unique, and the scattering problems (1.4.1)–(1.4.6) has a solution and this solution is unique.

Theorems 1.4.2 and 1.4.3 are proved. □

Let us now find analytical formulas for the solution of the scattering problems (1.4.1)–(1.4.6). It follows from formula (1.4.10) that

$$u(x) \sim u_0(x) + g(x, x_1) \left[\int_S e^{-ik\beta \cdot t} \sigma(t) dt + \kappa u(x_1)|D| \right],$$

$$|x - x_1| \gg a, \quad x_1 \in D. \tag{1.4.18}$$

To make this formula practically applicable, one needs to calculate the main terms of the asymptotic of the surface integral in (1.4.18) and of the term $u(x_1)$.

Note that

$$g(s, t) = g_0(s, t)[1 + O(ka)], \quad g_0(s, t) = \frac{1}{4\pi|s - t|}, \quad s, t \in S, \tag{1.4.19}$$

$$\frac{\partial g(s, t)}{\partial N_S} = \frac{\partial g_0}{\partial N_S}[1 + O(k^2 a^2)], \quad a \to 0, \tag{1.4.20}$$

$$A\sigma = 2 \int_S \frac{\partial g(s, t)}{\partial N_S} \sigma(t) dt \sim 2 \int_S \frac{\partial g_0(s, t)}{\partial N_S} \sigma(t) dt = A_0 \sigma, \tag{1.4.21}$$

$$Bu \sim \kappa \int_B g_0(x, y) u(y) dy = \kappa B_0 u, \tag{1.4.22}$$

$$B_1 u \sim B_{01} u, \quad B_{01} u = \kappa \int_B \frac{\partial}{\partial N_S} g(s, y) u(y) dy. \tag{1.4.23}$$

One has

$$\int_S e^{-ik\beta \cdot t}\sigma(t)dt \sim \int_S \sigma(t)dt - ik\beta_p \int_S t_p\sigma(t)dt. \qquad (1.4.24)$$

Let

$$Q := \int_S \sigma(t)dt, \quad Q_1 := Q - ik\beta_p \int_S t_p\sigma(t)dt. \qquad (1.4.25)$$

Then, by formula (1.4.18), one gets

$$u(x) \sim u_0(x) + g(x, x_1)[Q_1 + \kappa u(x_1)|D|], \quad |x - x_1| \gg a. \quad (1.4.26)$$

Let us derive analytical formula for Q_1 and $u(x_1)$. Integrate equation (1.4.15) over S and get

$$Q := \lambda \int_S A\sigma ds + 2\lambda \int_S B_1 u ds + 2\lambda \int_S u_{0N} ds. \qquad (1.4.27)$$

One has, by the divergence theorem,

$$2\lambda \int_S u_{0N} ds \sim 2\lambda \nabla^2 u_0(x_1)|D|. \qquad (1.4.28)$$

Furthermore, using formula (1.1.34), one gets:

$$\lambda \int_S A\sigma ds \sim \lambda \int_S A_0\sigma ds = -\lambda \int_S \sigma ds = -\lambda Q, \qquad (1.4.29)$$

and

$$2\lambda \int_S B_1 u ds = 2\lambda\kappa \int_D dx \nabla_x^2 \int_D g(x, y)u(y)dy.$$

Equation (1.4.12) implies

$$\int_D dx \nabla_x^2 \int_D g(x, y)u(y)dy = -k^2 \int_D dx \int_D g(x, y)u(y)dy - \int_D dx u(x)$$
$$\sim -u(x_1)|D|. \qquad (1.4.30)$$

Thus,

$$2\lambda \int_S B_1 u ds \sim -2\lambda\kappa u(x_1)|D|, \quad a \to 0. \qquad (1.4.31)$$

Therefore, equations (1.4.27)–(1.4.31) imply

$$Q \sim -\frac{2\lambda\kappa}{1+\lambda}u(x_1)|D| + \frac{2\lambda}{1+\lambda}\nabla^2 u_0(x_1)|D|. \tag{1.4.32}$$

In order to estimate $u(x_1)$ as $a \to 0$, let us integrate equation (1.4.10) over D and obtain

$$u(x_1)|D| \sim u_0(x_1)|D| + \int_D dx \int_S g(x,t)\sigma(t)dt$$

$$+ \kappa \int_D dx \int_D g(x,y)u(y)dy. \tag{1.4.33}$$

One has

$$\int_D dx \int_S g(x,t)\sigma(t)dt = \int_S dt\sigma(t)\int_D dxg(x,t) \sim QO(a^2), \tag{1.4.34}$$

and

$$\int_D dx \int_D g(x,y)u(y)dy = u(x_1)|D|O(a^2), \tag{1.4.35}$$

where we have used the relation

$$\int_D g(x,y)dy = O(a^2), \quad a \to 0, \tag{1.4.36}$$

which holds if $0.5\,\mathrm{diam}\,D \leq a$.

It follows from formulas (1.4.33)–(1.4.36) that

$$u(x_1)|D| \sim u_0(x_1)|D| + QO(a^2), \quad a \to 0, \tag{1.4.37}$$

where we have neglected the terms of higher order of smallness as $a \to 0$.

From (1.4.32) and (1.4.37), it follows that

$$Q \sim \frac{2\lambda}{1+\lambda}|D|\Big(-\kappa u_0(x_1) + \nabla^2 u_0(x_1)\Big), \quad a \to 0, \tag{1.4.38}$$

and

$$u(x_1) \sim u_0(x_1), \quad a \to 0 \tag{1.4.39}$$

Let us derive an analytical formula for the second integral in (1.4.25).

Multiply equation (1.4.15) by t_p and integrate over S. Take into account relation (1.4.29) and formulas (1.4.31) and (1.4.32) to get

$$2\lambda \int_S ds s_p B_1 u \sim O(a^4), \quad a \to 0. \tag{1.4.40}$$

Let us define now an analog of the matrix (1.3.6), as follows:

$$\beta_{pq}(\lambda) := \frac{1}{|D|} \int_S t_p \sigma_q(t) dt, \tag{1.4.41}$$

where the function $\sigma_q(t) := \sigma_q(t, \lambda)$ solves the equation

$$\sigma_q(t) = \lambda A \sigma_q(t) - 2\lambda N_q. \tag{1.4.42}$$

Since $|\lambda| = \left| \frac{1-\rho}{1+\rho} \right| < 1$, when $\rho > 0$, and the operator A has no eigenvalues in the interval $(-1, 1)$, equation (1.4.42) has a unique solution.

One has $u_{0N}(t) \sim u_{0N}(x_1), x_1 \in D$, and

$$2\lambda u_{0N} = -2\lambda N_q \left(-\frac{\partial u_0(x_1)}{\partial x_q} \right), \quad \sigma \sim -\sigma_q \frac{\partial u_0(x_1)}{\partial x_q}. \tag{1.4.43}$$

Therefore, neglecting the term (1.4.40), which is of higher order of smallness as $a \to 0$, one gets

$$\int_S t_p \sigma(t) dt = -|D| \beta_{pq}(\lambda) \frac{\partial u_0(x_1)}{\partial x_q}. \tag{1.4.44}$$

Consequently,

$$-ik\beta_p \int_S t_p \sigma(t) dt = ik\beta_{pq}(\lambda)\beta_p \frac{\partial u_0(x_1)}{\partial x_q} |D|, \tag{1.4.45}$$

and formulas (1.4.25), (1.4.38), and (1.4.45) yield the formula for Q_1 as follows:

$$Q_1 = \frac{2\lambda}{1+\lambda} |D| \left(\nabla^2 u_0(x_1) - \kappa u_0(x_1) \right) + ik\beta_{pq}(\lambda)\beta_p \frac{\partial u_0(x_1)}{\partial x_q} |D|, \tag{1.4.46}$$

where

$$\frac{2\lambda}{1+\lambda} = 1 - \rho, \quad \beta_p := \frac{(x - x_1)_p}{|x - x_1|}, \quad x_p := x \cdot e_p. \qquad (1.4.47)$$

From formulas (1.4.18), (1.4.26), (1.4.39), and (1.4.46), it follows that

$$u(x) \sim u_0(x) + g(x, x_1) \left[(1 - \rho)\Big(\nabla^2 u_0(x_1) - \kappa u_0(x_1) \Big) \right.$$

$$\left. + ik\beta_{pq}(\lambda)\frac{(x - x_1)_p}{|x - x_1|}\frac{\partial u_0(x_1)}{\partial x_q} + \kappa u_0(x_1) \right] |D|, \quad |x - x_1| \gg a.$$

$$(1.4.48)$$

Furthermore,

$$A(\beta, \alpha, k) = \frac{|D|}{4\pi} \left[(1 - \rho)\Big(\nabla^2 u_0(x_1) - \kappa u_0(x_1) \Big) \right.$$

$$\left. + ik\beta_{pq}(\lambda)\beta_p\frac{\partial u_0(x_1)}{\partial x_q} + \kappa u_0(x_1) \right]. \qquad (1.4.49)$$

Formulas (1.4.48) and (1.4.49) give our final result.
Note that $|A(\beta, \alpha, k)| = O(a^3)$ and the scattering is anisotropic.
Let us summarize the results.

Theorem 1.4.4. *If $ka \ll 1$ and $\rho > 0$, then the scattering problems (1.4.1)–(1.4.6) have a unique solution. This solution can be calculated by formula (1.4.48). The scattering amplitude is calculated by formula (1.4.49).*

1.5 Summary of the results

The results of this chapter can be summarized as follows.
Scattering problems (1.1.1)–(1.1.4) have a unique solution for any $\zeta, \text{Im}\zeta \leq 0$, including the limiting cases $\zeta = 0$ and $\zeta = \infty$.
The solution can be calculated by formula (1.1.44) if $ka \ll 1$ and the scattering amplitude, by formula (1.1.46).
One has $|A(\beta, \alpha, k)| = O(\zeta a^2)$ and the scattering is isotropic. If $\zeta = \infty$, then u can be calculated by formula (1.2.6) and the scattering

amplitude, by formula (1.2.7) for $ka \ll 1$. The scattering is isotropic and $|A(\beta, \alpha, k)| = O(a)$.

If $\zeta = 0$, then u can be calculated by formula (1.3.12), the scattering amplitude, by formula (1.3.14) if $ka \ll 1$, $|A(\beta, \alpha, k)| = O(a^3)$, and the scattering is anisotropic.

The scattering problems (1.4.1)–(1.4.6) with the interface (transmission) boundary condition has a unique solution if $\rho \geq 0$, $k^2 > 0$, $k_1^2 > 0$. If $ka \ll 1$, then this solution can be calculated by formula (1.4.48), the scattering amplitude, by formula (1.4.49), $|A(\beta, \alpha, k)| = O(a^3)$, and the scattering is anisotropic.

Chapter 2

Scalar Wave Scattering by Many Small Bodies of an Arbitrary Shape

2.1 Impedance bodies

Consider the many-body scattering problem in the case of small impedance bodies (particles) of an arbitrary shape:

$$(\nabla^2 + k^2)u = 0 \quad \text{in } D', \quad D' = R^3 \setminus D, \tag{2.1.1}$$

$$\text{where } D = \bigcup_{m=1}^{M} D_m, \quad a = \max_m \left(\frac{1}{2}\text{diam}D_m\right), \quad ka \ll 1,$$

$$\left.\frac{\partial u}{\partial N}\right|_{S_m} = \zeta_m u, \quad 1 \le m \le M, \quad \text{Im}\zeta_m \le 0, \tag{2.1.2}$$

$$u = u_0 + v, \tag{2.1.3}$$

$$\frac{\partial v}{\partial r} - ikv = o\left(\frac{1}{r}\right), \quad r = |x| \to \infty, \quad \frac{x}{|x|} := \beta := x^0. \tag{2.1.4}$$

The field v is the scattered field, u_0 is the incident field which satisfies equation (2.1.1) in $\mathbb{R}^3$. For example, the incident field is often the direction of the plane wave $u_0 = e^{ik\alpha \cdot x}$, $\alpha \in S^2$ describes the direction of the propagation of the incident wave, $k > 0$ is the wave number, N is the unit normal to $S := \bigcup_{m=1}^{M} S_m$ pointing into D', $S_m \in C^{1,2}$, $\lambda \in (0, 1]$.

Theorem 2.1.1. *Problems (2.1.1)–(2.1.4) have a unique solution.*

Proof. The proof is essentially the same as the proof of Theorem 1.1.1, and we leave the details to the reader.

Theorem 2.1.1 is proved. ☐

Let us look for the solution to problems (2.1.1)–(2.1.4) of the form

$$u(x) = u_0(x) + \sum_{m=1}^{M} \int_{S_m} g(x,t)\sigma_m(t)dt. \qquad (2.1.5)$$

This function solves problems (2.1.1)–(2.1.4) if and only if σ_m are chosen so that the boundary condition (2.1.2) are satisfied. These boundary conditions lead to a system of M integral equations. Since M can be very large, say, $M = 10^{12}$, it is not practically feasible to solve such a large system of simultaneous integral equations for unknown functions σ_m. By this reason we develop a new approach to solving problems (2.1.1)–(2.1.4). This approach uses essentially the assumption $ka \ll 1$. It consists in replacing the unfeasible task of finding M unknown numbers. This task is practically feasible and, moreover, it is justified physically.

Let us rewrite equation (2.1.5) as

$$u(x) = u_0(x) + \sum_{m=1}^{M} g(x, x_m)Q_m$$

$$+ \sum_{m=1}^{M} \int_{S_m} [g(x,t) - g(x,x_m)]\sigma_m(t)dt, \qquad (2.1.6)$$

where

$$Q_m := \int_{S_m} \sigma_m(t)dt, \quad x_m \in D_m. \qquad (2.1.7)$$

The numbers Q_m are the numbers we mentioned above. The central idea is to show that the second sum in (2.1.6) is negligible compared with the first as $a \to 0$. If this is done, then the solution to the scattering problems (2.1.1)–(2.1.4) is given by the formula

$$u(x) \sim u_0(x) + \sum_{m=1}^{M} g(x,x_m)Q_m, \qquad (2.1.8)$$

so that the scattering problem is solved if the numbers Q_m are found.

These numbers will be found from a linear algebra system. To derive this system, let us introduce the notion of the effective field acting on the jth small body. We denote this field $u_e(x)$ and define it as

$$u_e(x) := u_0(x) + \sum_{m \neq j}^{M} g(x, x_m) Q_m. \qquad (2.1.9)$$

Let us make the basic assumptions that hold through out the book for the many-body wave scattering in the case of the small bodies.

Assumption A: *The minimal distance d between neighboring bodies is much larger than the size a of the bodies:*

$$d \gg a. \qquad (2.1.10)$$

The number $\mathcal{N}(\Delta)$ of the small bodies in an arbitrary open set Δ is given by the following formula:

$$\mathcal{N}(\Delta) = \frac{1}{a^{2-\kappa}} \int_\Delta N(x) dx [1 + o(1)], \quad a \to 0, \qquad (2.1.11)$$

where $N(x) \geq 0$ is a continuous function, $\kappa \in [0, 1)$ is a number,

$$\zeta_m = \frac{h(x_m)}{a^\kappa}, \quad \mathrm{Im} h(x) \leq 0, \qquad (2.1.12)$$

and $h(x)$ is a continuous function, $1 \leq m \leq M$.

    The parameter κ, the function h, and the function $N(x) \geq 0$ can be chosen by the experimenter as he (she) wishes.

    The assumption $\mathrm{Im} h \leq 0$ guarantees by Theorem 2.1.1 the uniqueness of the solution to problems (2.1.1)–(2.1.4). The impedance function ζ does not have to depend on a as in (2.1.12), but no physical restrictions prevent such a dependence. We will see in Chapter 3 that such a dependence can be used practically. We assume that the small bodies are distributed according to the law (2.1.11) in an arbitrary finite region Ω, so that $N(x) = 0$ in $\Omega' = \mathbb{R}^3 \setminus \Omega$.

    Assumption (2.1.10) allows one to consider jth body as the body placed in the exterior field $u_e(x)$ defined by formula (2.1.9). We do not put index j on $u_e(x) := u_{e,j}(x)$ to keep the notations simpler.

Let us denote

$$u_e(x_j) := u_j, \quad 1 \le j \le M. \tag{2.1.13}$$

Our first task is to express Q_m in terms of u_m. This is done by formula (1.1.42) in which $u_0(x)$ is replaced by $u_e(x)$, as follows:

$$Q_m = -\zeta_m |S_m| u_{e,m}(x_m) := -\zeta_m |S_m| u_m, \tag{2.1.14}$$

and we use the notation $u_m := u_{e,m}$ for simplicity.

If

$$S_m = c_m a^2, \quad c_m = c = \text{const},$$

then Assumption (2.1.12) and formula (2.1.14) yield

$$Q_m = -c_m h_m a^{2-\kappa} u_m, \quad 1 \le m \le M, \quad h_m := h(x_m). \tag{2.1.15}$$

From (2.1.15) and (2.1.9), it follows that

$$u_j = u_{0j} - \sum_{\substack{m=1 \\ m \ne j}}^{M} g_{jm} c_m h_m a^{2-\kappa} u_m, \quad 1 \le j \le M, \tag{2.1.16}$$

where

$$g_{jm} := g(x_j, x_m), \quad u_{0j} := u_0(x_j). \tag{2.1.17}$$

Linear algebraic system (2.1.16) (LAS) allows one to find numbers $u_m, 1 \le m \le M$. If these numbers are found, then the numbers Q_m are found by formula (2.1.15) and the solution to the scattering problems (2.1.1)–(2.1.4) is found by formula (2.1.8).

Let us prove that the neglected term in formula (2.1.6) is much smaller than the term we kept.

Lemma 2.1.1. *One has*

$$J_1 := |g(x, x_m)Q_m| \gg \left| \int_{S_m} [g(x, t) - g(x, x_m)]\sigma_m(t)dt \right| := J_2, \quad a \to 0. \tag{2.1.18}$$

Proof.   By (2.1.10), one has

$$|g(x, x_m)Q_m| \geq \frac{Q_m}{4\pi d}, \quad |x - x_m| \geq d. \qquad (2.1.19)$$

Since $|x - x_m| \geq d \gg a$ and $|t - x_m| \leq a$, one has

$$|g(x, t) - g(x, x_m)| \leq \max\left\{ O\left(\frac{a}{d^2}\right), O\left(\frac{ka}{d}\right) \right\}, \qquad (2.1.20)$$

and

$$\frac{J_2}{J_1} \leq O\left(\frac{a}{d} + ka\right) \ll 1. \qquad (2.1.21)$$

Lemma 2.1.1 is proved. $\square$

Our next task is to reduce the order of the linear algebraic system (2.1.16). For practical calculations this is an important task. Let us explain how to do this.

Let us partition the domain Ω, where the small bodies are distributed, into a union of P cubes, Δ_p. These cubes do not have common interior points but may have common parts of the boundary. Let $x_p \in \Delta_p$. We assume that the side of Δ_p is b and the following conditions hold:

$$a \ll d \ll b. \qquad (2.1.22)$$

One assumes that $d = d(a)$ and $b = b(a)$. Then, the assumption, which implies (2.1.22), can be written as

$$\lim_{a \to 0} \frac{a}{d} = 0, \quad \lim_{a \to 0} \frac{d}{b} = 0. \qquad (2.1.23)$$

One can rewrite (2.1.16) as

$$u_q = u_{0q} - \sum_{p \neq q}^{P} c_p g_{qp} h_p u_p a^{2-\kappa} \sum_{x_m \in \Delta_p} 1, \quad 1 \leq q \leq P. \qquad (2.1.24)$$

If $\mathrm{diam}\Delta_p \ll 1$ and $N(x)$ is continuous, then (see formula (2.1.11))

$$a^{2-\kappa} \mathcal{N}(\Delta_p) = N(x_p)|\Delta_p|[1 + o(1)], \quad a \to 0. \qquad (2.1.25)$$

It follows from (2.1.24) and (2.1.25) that, assuming $c_p = c =$ const, $|\Delta_p| = b^3$ is the volume of Δ_p, one has

$$u_q = u_{0q} - c \sum_{\substack{p \neq q}}^{P} g_{qp} h_p N_p u_p |\Delta_p|, \quad 1 \leq q \leq P, \qquad (2.1.26)$$

where the $o(1)$ term in (2.1.25) is neglected and the continuity of the functions $h(x), N(x)$ and $u(x)$ allows one to replace h_m and u_m by h_p and u_p for any $x_m \in \Delta_p$, and to use the formula

$$\int_{\Delta_p} N(x)dx \sim N(x_p)|\Delta_p|, \quad a \to 0.$$

Equation (2.1.26) is the linear algebraic system (LAS2) of the order $P \ll M$.

This concludes the description of the process of the reduction of the order of LAS (2.1.16).

One can see that LAS2 (2.1.26) is obtained if one uses the collocation method (see Chapter 12) for solving the integral equation

$$u(x) = u_0(x) - c \int_{\Omega} g(x, y)h(y)N(y)u(y)dy. \qquad (2.1.27)$$

Integral equation (2.1.27) is the equation for the limiting effective field in the medium in which many small particles are embedded when the size of a particle tends to zero, $a \to 0$, while the number of the particles $\mathcal{N}(D) \to \infty$ according to the law (2.1.11), that is $\mathcal{N}(D) = O\left(\frac{1}{a^{2-\kappa}}\right)$.

Lemma 2.1.2. *If $Imq \leq 0$ and q is compactly supported, then equation (2.1.27) has a unique solution.*

Proof. Indeed, it is a Fredholm-type equation, that is, an equation for which the Fredholm alternative is valid (see Appendix A). Moreover, the homogeneous version

$$u(x) = -c \int_{\Omega} g(x, y)h(y)N(y)u(y)dy$$

of this equation has only the trivial solution. To prove this, apply the operator $\nabla^2 + k^2$ to the above equation and get

$$[\nabla^2 + k^2 - cN(x)h(x)]u = 0 \quad \text{in } \mathbb{R}^3. \qquad (2.1.28)$$

This is a Schrödinger equation with a compactly supported potential $q(x) = cN(x)h(x)$, $q(x) = 0$ in Ω'. If $\text{Im}\,h \leq 0$, then the solution to equation (2.1.28) satisfying the radiation condition (2.1.4) must be identically equal to zero. To prove this, multiply equation (2.1.28) by $\bar{u}$, subtract the complex conjugate of equation (2.1.28) multiplied by u, integrate over the ball $B_R := \{x : |x| \leq R\}$ of large radius R, and use the radiation condition. The result is

$$\lim_{R \to \infty} \left(2ik \int_{|s|=R} |u|^2 ds - 2i \int_{B_R} \text{Im}\,q(x)|u|^2 dx \right) = 0. \qquad (2.1.29)$$

If $\text{Im}\,q \leq 0$, then one can concludes that

$$\lim_{R \to \infty} \int_{|s|=R} |u|^2 ds = 0. \qquad (2.1.30)$$

This and equation (2.1.28) imply that $u = 0$ in Ω', and, by the unique continuation theorem for solutions to elliptic equation (2.1.28), it follow that $u = 0$ in $\mathbb{R}^3$. Lemma 2.1.2 is proved. $\square$

Consequently, by the Fredholm alternative, the operator $(I+T)^{-1}$ is bounded, where

$$Tu := c \int_\Omega g(x,y)h(y)N(y)u(y)dy.$$

Remark 2.1.1. Let us denote $\eta := \text{Im}\,h(x)$. The operator $T = T_\eta$ depends continuously on the parameter η in the norm of the operators in the space $C(\Omega)$. Since the operator $(I + T_\eta)^{-1}$ is bounded for $\text{Im}\,\eta \leq 0$ and depends on η continuously in the norm of the operators, it must be bounded for sufficiently small positive $\eta > 0$. For such η, one has $\arg(-cN(x)h(x)) < 0$.

Convergence of the collocation method for solving integral equation (2.1.27) is studied in detail in Chapter 12, where a one-to-one correspondence between the function that solves equation (2.1.27) approximately with any desired accuracy and the vector $(u_1, \ldots, u_P)$

that solves LAS (2.1.26) is established. This convergence justifies the transition in the limit $a \to 0$ from the LAS (2.1.26) to the integral equation (2.1.27). In this sense it plays the role of the theory (see Bensoussan *et al.* (2011); Marčenko and Khruslov (2006); Zhikov *et al.* (1994)), but our theory does not require the periodicity assumption, used in the theory, the spectrum of our problems (2.1.1)–(2.1.4) are continuous and not discrete, as in the usual theory, and our operator is non-self-adjoint, also in contrast to the standard theory.

The dependence of c_p on p in formula (2.1.24) can describe the particles which are not identical. For simplicity, we have assumed in (2.1.26) that $c_p = c$ does not depend on p, that is, that all the particles are identical.

Let us apply the operator $\nabla^2 + k^2$ to equation (2.1.27) and use the known equation $(\nabla^2 + k^2)g(x, y) = -\delta(x - y)$ to get equation (2.1.28). This equation can be interpreted physically as follows: the limiting medium, obtained by embedding many small particles according to the distribution law (2.1.11) with the boundary impedances defined in (2.1.12), has a new refraction coefficient defined by the relation

$$n^2(x) = 1 - ck^{-2}h(x)N(x), \tag{2.1.31}$$

so

$$n(x) = [1 - ck^{-2}h(x)N(x)]^{1/2}. \tag{2.1.32}$$

Since the functions $h(x), \mathrm{Im}h(x) \leq 0$, and $N(x) \geq 0$, can be chosen by an experimenter formula (2.1.32) gives very wide possibilities for creating materials with a desired refraction coefficient by embedding into a given material many small impedance particles according to the distribution law (2.1.11) with boundary impedances ζ_m defined in (2.1.12).

This is discussed in Chapter 3.

Let us summarize the results.

Theorem 2.1.2. *Problems* (2.1.1)–(2.1.4) *have a unique solution which can be calculated by formula* (2.1.8) *in which the numbers* Q_m *are given by formula* (2.1.14) *and the numbers* u_m *are found from the linear algebraic system* (2.1.16). *As* $a \to 0$ *and Assumptions* (2.1.10)–(2.1.12), (2.1.22) *hold, then the field tends to the (unique) solution of the integral equation* (2.1.27). *The resulting limiting medium has the refraction coefficient defined in* (2.1.32).

2.2   The Dirichlet boundary condition

Let us consider the scattering problems (2.1.1)–(2.1.4) with the impedance boundary condition (2.1.2) replaced by the Dirichlet boundary condition

$$u|_{S_m} = 0, \quad 1 \le m \le M. \tag{2.2.1}$$

Theorem 2.2.1. *Problems* (2.1.1), (2.2.1), (2.1.3), *and* (2.1.4) *have a unique solution.*

Proof.   Let us first prove that the above problem has no more than one solution. This is equivalent to proving that the corresponding homogeneous problem, that is the problem with $u_0 = 0$, has only the trivial solution. Let u be a solution to the homogeneous problems (2.1.1), (2.2.1), (2.1.3), and (2.1.4). Multiply equation (2.1.1) by $\bar{u}$, integrate over $D'_R := D' \cap B_R$, and use Green's formula to get

$$0 = \int_{D'_R} (k^2 |u|^2) dx + ik \int_{S'_R} |u|^2 ds + o(1), \quad R \to \infty. \tag{2.2.2}$$

Taking the imaginary part of (2.2.2), one obtains

$$\lim_{R \to \infty} \int_{S_R} |u|^2 ds = 0. \tag{2.2.3}$$

This and the equation (2.1.1) imply that $u = 0$ in D'. Thus, it is proved that problems (2.1.1), (2.2.1), (2.1.3), and (2.1.4) have no more than one solution.

Let us prove the existence of the solution of the form (2.1.5). It is sufficient to prove that σ_m are uniquely defined by the boundary condition (2.2.1) as follows:

$$\sum_{m=1}^{M} \int_{S_m} g(s,t)\sigma_m(t) dt = -u_0(s), \quad s \in S_j, \quad 1 \le j \le M. \tag{2.2.4}$$

Let us rewrite this equation as

$$\sigma_j + \sum_{m=1}^{M} T_{jj}^{-1} T_{jm}\sigma_m = -T_{jj}^{-1} u_0, \quad 1 \le j \le M, \tag{2.2.5}$$

where

$$T_{jm}\sigma_m = \int_{S_m} g(s,t)\sigma_m(t)dt, \quad s \in S_j.$$

Clearly, $T_{jm} : L^2(S_m) \to C^\infty(S_j)$, if $j \neq m$. It is known that $T_{jj} : L^2(S_j) \to H^1(S_j)$ is an isomorphism if k^2 is not an eigenvalue of the Dirichlet Laplacian in D_j. Since D_j is small, this condition is satisfied. Therefore, $T_{jj}^{-1} : H^1(S_j) \to L^2(S_j)$ is a bounded operator and $T_{jj}^{-1}T_{jm}$ is a compact operator from $L^2(S_m)$ into $L^2(S_j)$. Thus, equation (2.2.5) is of Fredholm type. It is uniquely solvable because the corresponding homogeneous equation has only the trivial solution, as follows from the uniqueness result proved above.

Theorem 2.2.1 is proved.   □

Let us derive a formula for the solution. Formula (2.1.5) implies, as in Section 2.1, that formula (2.1.9) holds. The numbers Q_m now can be found by a formula similar to (1.2.5):

$$Q_m = -C_m u_e(x_m) := -C_m u_m, \qquad (2.2.6)$$

where C_m is the electrical capacitance of a perfect conductor with the shape D_m, and $u_e(x_m) := u_m$ is the value of the effective field, defined by formula (2.1.9), and the point $x_m \in D_m$. The numbers u_m can be calculated by solving a linear algebraic system (LAS) similar to (2.1.16):

$$u_j = u_{0j} - \sum_{m \neq j}^{M} g_{jm}C_m u_m, \quad 1 \le m \le M. \qquad (2.2.7)$$

The LAS of the reduced order P, analogous to (2.1.26), is of the form

$$u_q = u_{0q} - \sum_{p \neq q}^{P} g_{pq}C_p a^{-1}N_p u_p |\Delta_p|, \quad 1 \le p \le P. \qquad (2.2.8)$$

Since $C_m = O(a)$, the distribution law, analogous to (2.1.11), takes now the form:

$$\mathcal{N}(\Delta) = \frac{1}{a} \int_\Delta N(x)dx[1 + o(1)], \quad a \to 0. \qquad (2.2.9)$$

Assume for simplicity that all the small bodies are identical. Then

$$C_p a^{-1} = c, \quad c = \text{const} > 0. \qquad (2.2.10)$$

Passing to the limit $a \to 0$ in equation (2.2.8) yields the integral equation for the limiting effective field

$$u(x) = u_0(x) - c \int_\Omega g(x, y) N(y) u(y) dy. \qquad (2.2.11)$$

This equation is an analog of equation (2.1.27).

Applying the operator $\nabla^2 + k^2$ to equation (2.2.11), one gets

$$[\nabla^2 + k^2 - cN(x)]u = 0 \quad \text{in } \mathbb{R}^3. \qquad (2.2.12)$$

This means that

$$n(x) = [1 - k^{-2} cN(x)]^{1/2}. \qquad (2.2.13)$$

This is a formula analogous to (2.1.32).

Let us summarize the results.

Theorem 2.2.2. *Assume* (2.2.9), (2.1.10), *and* (2.2.10). *Then the solution to problems* (2.1.1), (2.2.1), (2.1.3), *and* (2.1.4) *exists, is unique, can be calculated by formula* (2.2.8), *where* Q_m *are calculated by formula* (2.2.6) *and* u_m *solve* (2.2.7).

As $a \to 0$, *the limiting effective field solves equation* (2.2.11). *The limiting medium has refraction coefficient* (2.2.13).

2.3 The Neumann boundary condition

Let us now consider the scattering problems (2.1.1)–(2.1.4) with the boundary condition (2.1.2) replaced by the Neumann boundary condition

$$u_N|_{S_m} = 0, \quad 1 \le m \le M. \qquad (2.3.1)$$

Theorem 2.3.1. *The aforementioned problem has a unique solution. This solution is of the form* (2.1.5). *The function* $\sigma_m(t)$ *in formula* (2.1.5) *can be uniquely found from the boundary conditions* (2.3.1).

Proof.   The uniqueness and existence of the solution to problems
(2.1.1), (2.3.1), (2.1.3), and (2.1.4) follow from Theorem 2.1.1 because
$\zeta = 0$ yields this problem. The solution of the form (2.1.5) can be
found: if one substitutes (2.1.5) into (2.3.1), one gets a Fredholm-type
system of integral equations

$$\frac{A_j\sigma_j - \sigma_j}{2} + \sum_{m\neq j}^{M} \frac{\partial T_{jm}\sigma_m}{\partial N_s} = -u_{0N}(s), \quad s \in S_j, \qquad (2.3.2)$$

where

$$T_{jm}\sigma_m := \int_{S_m} \frac{\partial g(s,t)}{\partial N_S}\sigma_m(t)dt, \quad s \in S_j. \qquad (2.3.3)$$

Since the homogeneous equation (2.3.2) has only the trivial solution,
the Fredholm-type system of equation (2.3.2) is uniquely solvable.
   Theorem 2.3.1 is proved.   □

The solution u to problems (2.1.1), (2.3.1), (2.1.3), and (2.1.4) can
be calculated by formula (2.1.8) where the numbers Q_m can be calculated by the formula analogous to (1.3.11) as follows:

$$Q_m \sim |D_m| \left(\nabla^2 u_e(x_m) + ik\beta_{pq}(x_m)\frac{(x-x_m)_p}{|x-x_m|}\frac{\partial u_e(x_m)}{\partial x_q} \right), \quad (2.3.4)$$

where $\beta_p := \beta_p(x_m) = \frac{(x-x_m)_p}{|x-x_m|}$, $(x_m)_p$ is the pth component of the
vector x_m, and over the repeated indices p and q one sums up.
   Let us assume that the following limits exists:

$$\lim_{a\to 0} \frac{\sum_{x_m\in\Delta_\nu} |D_m|}{|\Delta_\nu|} = \rho(x), \quad x \in \Delta_\nu \qquad (2.3.5)$$

$$\lim_{a\to 0} \frac{\sum_{x_m\in\Delta_\nu} \beta_{pq}(x_m)|D_m|}{|\Delta_\nu|} = B_{pq}(x), \quad x \in \Delta_\nu. \qquad (2.3.6)$$

Here, Δ_ν is a partition cube, $|\Delta_\nu|$ is its volume, the side of Δ_ν is
equal to b, and assumptions (2.1.22) and (2.1.23) hold. The function
$\rho(x) > 0$ and $B_{pq}(x)$ are assumed to be smooth.

If these assumptions hold, then the effective field

$$u_e(x) = u_0(x) + \sum_{m \neq j} g(x, x_m) \left(\nabla^2 u_e(x_m) + ik\beta_{pq}(x_m) \right.$$

$$\left. \times \frac{(x - x_m)_p}{|x - x_m|} \frac{\partial u_e(x_m)}{\partial x_q} \right) |D_m| \qquad (2.3.7)$$

has a limit $u(x)$ as $a \to 0$, and this limit satisfies the following equation:

$$u(x) = u_0(x) + \int_\Omega g(x, y) \left(\rho(y) \nabla^2 u(y) \right.$$

$$\left. + ik B_{pq}(y) \frac{(x - y)_p}{|x - y|} \frac{\partial u(y)}{\partial y_q} \right) dy. \qquad (2.3.8)$$

This is an integral-differential equation which is not equivalent to a local differential equation in contrast to the scattering problems considered in Sections 2.1 and 2.2.

Let us check that equation (2.3.8) is of Fredholm type. Apply the operator $\nabla^2 + k^2$ to equation (2.3.8) and use the equation

$$(\nabla^2 + k^2) g(x, y) = -\delta(x - y) \qquad (2.3.9)$$

to get

$$(\nabla^2 + k^2) u = -\rho(x) \nabla^2 u(x) + ik(\nabla^2 + k^2)$$

$$\times \int_\Omega g(x, y) B_{pq}(y) \frac{(x - y)_p}{|x - y|} \frac{\partial u(y)}{\partial y_q} dy. \qquad (2.3.10)$$

This equation can be rewritten as

$$\nabla^2 u + \frac{k^2 u}{1 + \rho(x)} = \frac{ik}{1 + \rho(x)} (\nabla^2 + k^2)$$

$$\times \int_\Omega g(x, y) \frac{(x - y)_p}{|x - y|} B_{pq}(y) \frac{\partial u(y)}{\partial y_q} dy. \qquad (2.3.11)$$

Let us calculate the expression

$$I := (\nabla_x^2 + k^2) \left(g(x, y) \frac{(x - y)_p}{|x - y|} \right)$$

in the sense of distributions. One has

$$I = -\delta(x-y)\frac{(x-y)_p}{|x-y|} + 2\nabla_x g \cdot \nabla \frac{(x-y)_p}{|x-y|} + g\nabla^2 \frac{(x-y)_p}{|x-y|},$$

(2.3.12)

and

$$\delta(x-y)\frac{(x-y)_p}{|x-y|} = 0,$$

(2.3.13)

$$\nabla_x g \cdot \nabla \frac{(x-y)_p}{|x-y|} = g' \frac{x-y}{|x-y|} \cdot \left(\frac{e_p}{|x-y|} - \frac{(x-y)_p(x-y)}{|x-y|^3} \right) = 0,$$

(2.3.14)

$$g' = \frac{e^{ik|x-y|}}{4\pi|x-y|} \left(ik - \frac{1}{|x-y|} \right).$$

(2.3.15)

Finally,

$$g\nabla^2 \frac{(x-y)_p}{|x-y|} = g\left(-4\pi\delta(x-y)(x-y)_p - 2e_p \frac{x-y}{|x-y|^3} \right)$$

$$= -2g\frac{x_p - y_p}{|x-y|^3}.$$

(2.3.16)

Thus, in the distributional sense, one has:

$$I = -2g\frac{x_p - y_p}{|x-y|^3}.$$

(2.3.17)

Therefore, formula (2.3.10) can be written as

$$\nabla^2 u + \frac{k^2 u}{1+\rho(x)} = -\frac{2ik}{1+\rho(x)} \int_\Omega \frac{e^{ik|x-y|}(x_p - y_p)}{4\pi|x-y|^4}$$

$$\times g(x,y)B_{pq}(y)\frac{\partial u(y)}{\partial y_q} dy.$$

(2.3.18)

The integral in (2.3.18) is a singular integral operator bounded in $L^2(\Omega)$ by Theorem X.5.1 in Mikhlin and Prössdorf (1986), p. 257. Therefore, the integral operator in (2.3.18) maps bounded in $L^2(\Omega)$ sets $\frac{\partial u}{\partial y_q}$ into bounded in $L^2(\Omega)$ sets. Since the left-hand side of (2.3.18) is an elliptic operator of the second order and the integral operator in (2.3.18) is bounded as an operator from $H^1(\Omega)$

into $L^2(\Omega)$, it follows that equation (2.3.18) is of Fredholm type. By $H^l(\Omega) = W^{l,2}(\Omega)$, the Sobolev space is denoted, see Gilbarg and Trudinger (1983), and the Appendix A.

2.4   The transmission boundary condition

Consider now the wave scattering problem with the following transmission boundary condition:

$$(\nabla^2 + k^2)u = 0 \quad \text{in } D' = \mathbb{R}^3 \setminus D, \quad D = \bigcup_{m=1}^{M} D_m, \qquad (2.4.1)$$

$$(\nabla^2 + k_m^2)u = 0 \quad \text{in } D_m, \qquad (2.4.2)$$

$$u^+ = u^-, \quad \rho_m u_N^+ = u_N^- \quad \text{on } S_m, \quad 1 \le m \le M, \qquad (2.4.3)$$

$$u = u_0 + v, \quad \frac{\partial v}{\partial r} - ikv = o\left(\frac{1}{r}\right), \quad r = |x| \to \infty, \qquad (2.4.4)$$

$$(\nabla^2 + k^2)u_0 = 0, \quad u_0 = e^{ik\alpha \cdot x}, \quad \alpha \in S^2. \qquad (2.4.5)$$

We assume $k^2 > 0$, $\mathrm{Im}k_m^2 \ge 0$, $\rho_m > 0$. As in Section 1.4, one proves the uniqueness and existence theorem.

Theorem 2.4.1. *Problems* (2.4.1)–(2.4.5) *have a solution.*

Proof.   The proof is analogous to the proof of Theorems 1.4.1 and 1.4.2. Details are left to the reader, see also Ramm (1986).
   Theorem 2.4.1 is proved.   □

The solution to the many-body scattering problems (2.4.1)–(2.4.5) can be found of the form similar to (1.4.10):

$$u(x) = u_0(x) + \sum_{m=1}^{M} \int_{S_m} g(x,t)\sigma_m(t)dt + \sum_{m=1}^{M} \kappa_m \int_{D_m} g(x,y)u(y)dy,$$
$$\tag{2.4.6}$$

where

$$\kappa_m := k_m^2 - k^2.$$

One verifies, as in Section 1.4, that the function (2.4.6) solves equations (2.4.1), (2.4.2), (2.4.4) and the first equation (2.4.3). Thus, it solves problems (2.4.1)–(2.4.5) if it satisfies the second equation (2.4.3). These equations can be written similarly to equations (1.4.13) and (1.4.15), as follows:

$$
\rho_j \frac{A_j \sigma_j + \sigma_j}{2} - \frac{A_j \sigma_j - \sigma_j}{2} + \sum_{m \neq j}^{M} \int_{S_m} (\rho_j - 1) \frac{\partial g(s,t)}{\partial N} \sigma_m(t) dt
$$

$$
+ \sum_{m=1}^{M} \kappa_m (\rho_j - 1) \frac{\partial}{\partial N} \int_{D_m} g(x,y) u(y) dy + (\rho_j - 1) u_{0N} = 0.
$$

$$(2.4.7)$$

One can use formula (1.4.48) to write the effective (self-consistent) field acting on the jth small body

$$
u_e(x) = u_0(x) + \sum_{m \neq j}^{M} g(x, x_m) \left((1 - \rho_m)[\nabla^2 u_e(x_m) - \kappa_m u_e(x_m)] \right.
$$

$$
\left. + i k \beta_{pq}(x_m) \frac{\partial u_e}{\partial x_q} \frac{(x - x_m)_p}{|x - x_m|} \right) |D_m|
$$

$$
+ \sum_{m \neq j}^{M} g(x, x_m) \kappa_m u_e(x_m) |D_m|, \quad |x - x_j| \sim a. \qquad (2.4.8)
$$

Let us derive the integral equation for the limit of the effective field as $a \to 0$ while the assumptions (2.1.22), (2.1.23) hold.

We assume also that

$$
\lim_{a \to 0} \frac{\sum_{x_m \in \Delta_p} \rho_m |D_m|}{|\Delta_p|} = \rho(x), \quad x \in \Delta_p, \qquad (2.4.9)
$$

$$
\lim_{a \to 0} \frac{\sum_{x_m \in \Delta_p} \beta_{pq}(x_m)|D_m|}{|\Delta_p|} = B_{pq}(x), \quad x \in \Delta_p, \qquad (2.4.10)
$$

$$
\lim_{a \to 0} \frac{\sum_{x_m \in \Delta_p} \kappa_m |D_m|}{|\Delta_p|} = K(x), \qquad (2.4.11)
$$

and the functions $\rho(x) \geq 0$, $B_{pq}(x)$ and $K(x)$ are smooth. Then the limiting equation for $u(x) = \lim_{a \to 0} u_e(x)$ is

$$u(x) = u_0(x) + \int_{\Omega} g(x, y) \Big[(1 - \rho(y))(\nabla^2 u(y) - K(y)u(y))$$

$$+ ik B_{pq}(y) \frac{\partial u(y)}{\partial y_q} \frac{x_p - y_p}{|x - y|} + K(y)u(y) \Big] dy. \tag{2.4.12}$$

2.5   Wave scattering in an inhomogeneous medium

It is both of theoretical and practical interest to consider wave scattering by one and many small bodies of an arbitrary shape embedded into an inhomogeneous medium. Let us consider this problem for impedance particles. One wants to find the solution to the following problem:

$$[\nabla^2 + k^2 n_0^2(x)]u = 0 \quad \text{in } D' = \mathbb{R}^3 \setminus \bigcup_{m=1}^{M} D_m, \tag{2.5.1}$$

$$\frac{\partial u}{\partial N} = \zeta_m u \quad \text{on } S_m, \quad 1 \leq m \leq M, \quad \operatorname{Im}\zeta_m \leq 0, \tag{2.5.2}$$

$$u = u_0 + v, \quad \frac{\partial v}{\partial r} - ikv = o\left(\frac{1}{r}\right), \quad r = |x| \to \infty. \tag{2.5.3}$$

Let us assume that

$$[\nabla^2 + k^2 n_0^2(x)]u_0 = 0 \quad x \in \mathbb{R}^3,$$

the function $n_0^2(x)$ is given and satisfies the conditions:

$$n_0^2(x) = 1, \quad x \in \Omega' = \mathbb{R}^3 \setminus \Omega, \tag{2.5.4}$$

Ω is a bounded domain where the small bodies are distributed,

$$D := \bigcup_{m=1}^{M} D_m \subset \Omega.$$

Inside Ω, the function $n_0^2(x)$ is an arbitrary continuous function, satisfying the condition

$$\operatorname{Im} n_0^2(x) \geq 0, \tag{2.5.5}$$

which physically means that the medium is not active, that is, it does not generate the energy.

Let $G(x, y)$ be the Green's function satisfying the equation

$$[\nabla^2 + k^2 n_0^2(x)]G(x, y) = -\delta(x - y) \quad \text{in } \mathbb{R}^3, \quad k^2 = \text{const} > 0, \tag{2.5.6}$$

and the radiation condition

$$\frac{\partial G}{\partial |x|} - ikG = o\left(\frac{1}{|x|}\right), \quad |x| \to \infty. \tag{2.5.7}$$

Since $n_0^2(x)$ is known, one may assume that $G(x, y)$ is a known function. Of course, if $n_0^2(x)$ is arbitrary, then one cannot calculate $G(x, y)$ in a closed form, but one can calculate it numerically.

Let us look for a solution to problems (2.5.1)–(2.5.5) of the form

$$u(x) = u_0(x) + \sum_{m=1}^{M} \int_{S_m} G(x, t)\sigma_m(t)dt. \tag{2.5.8}$$

As in Section 2.1, one proves that problems (2.5.1)–(2.5.5) have a unique solution, and this solution can be found as in (2.5.8). Assume that conditions (2.1.11), (2.1.12), (2.1.22), and (2.1.23) hold. Then one can repeat all the arguments given in Section 2.1 and obtain the results similar to Theorems 2.1.1, 2.1.2 and Lemma 2.1.1.

Note that the asymptotic of the function $G(x, y)$ as $|x - y| \to 0$ is the same as that of $g(x, y)$, and estimates of $G(x, y)$ as $|x-y| \to \infty$ are similar to those of $g(x, y) = \frac{e^{ik|x-y|}}{4\pi|x-y|}$. Let us explain these statements briefly.

One has the following equation for $G(x, y)$:

$$G(x, y) = g(x, y) + \int_{\Omega} g(x, z)q(z)G(z, y)dz, \tag{2.5.9}$$

where

$$q(z) := k^2[n_0^2(z) - 1], \quad q(z) = 0, \quad z \in \Omega'. \tag{2.5.10}$$

Note that

$$g(x, y) = g_0(x, y)[1 + O(|x - y|)], \quad |x - y| \to 0;$$

$$g_0(x, y) = \frac{1}{4\pi|x - y|}. \tag{2.5.11}$$

Equation (2.5.9) is of Fredholm type and its homogeneous version has only the trivial solution (we prove this later), so equation (2.5.9) has a unique solution. The integral in (2.5.9) is a bounded function. Therefore,

$$G(x,y) = g_0(x,y)[1 + O(|x-y|)], \quad |x-y| \to 0. \qquad (2.5.12)$$

The solution to (2.5.9) clearly satisfies the radiation condition and the estimate, as follows:

$$|G(x,y)| \le \frac{c}{|x|}, \quad |x| \to \infty, \quad y \in \Omega. \qquad (2.5.13)$$

Let us now prove that the homogeneous version of (2.5.9) has only the trivial solution in $C(\Omega)$. Suppose that

$$w(x) = \int_\Omega g(x,z)q(z)w(z)dz. \qquad (2.5.14)$$

Then w satisfies the radiation condition and solves the equation

$$(\nabla^2 + k^2 + q(x))w = 0, \quad k^2 > 0, \quad \mathrm{Im}\, q \ge 0, \qquad (2.5.15)$$

where $q(x)$ has compact support.

Multiply equation (2.5.15) by $\overline{w}$ and the complex conjugate of (2.5.15) by w, subtract from the first equation the second, integrate the result over the ball $B_R := \{x : |x| \le R\}$, use Green's formula and the radiation condition for w, and get the following:

$$\lim_{R\to\infty} \left(2ik \int_{|s|=R} |w|^2 ds + 2i \int_\Omega \mathrm{Im}\, q(x)|w|^2 dx \right) = 0. \qquad (2.5.16)$$

Since $\mathrm{Im}\, q \ge 0$, it follows from (2.5.16) that

$$\lim_{R\to\infty} \int_{|s|=R} |u|^2 ds = 0. \qquad (2.5.17)$$

Consequently, by the result we have used several times already, $w = 0$ in Ω'. This and the unique continuation theorem for solutions to elliptic equation (2.5.15) imply that $w = 0$. This completes the proof.

Let us summarize our result.

Theorem 2.5.1. *Problems* (2.5.1)–(2.5.5) *have a unique solution. This solution is of the form* (2.5.8). *The effective field in the medium is defined by the formula*

$$u_e(x) = u_0(x) + \sum_{m \neq j} G(x, x_m) Q_m. \qquad (2.5.18)$$

The numbers Q_m are given in formula (2.1.14) *and the numbers u_m in this formula are found by solving the linear algebraic system*

$$u_j = u_{0j} - \sum_{m \neq j} G_{jm} c_m h_m a^{2-\kappa} u_m, \quad 1 \leq j \leq M, \qquad (2.5.19)$$

where $G_{jm} := G(x_j, x_m)$, and assumptions (2.1.10)–(2.1.12) *hold. If $a \to 0$ and assumptions* (2.1.22)–(2.1.23) *hold, then the effective field has a limit $u(x)$, which is the unique solution of the equation*

$$u(x) = u_0(x) - c \int_\Omega G(x, y) h(y) N(y) u(y) dy \qquad (2.5.20)$$

where $c > 0$ is a constant defined by the equation $|S| = ca^2$, if one assumes that the small bodies D_m are identical.

The physical interpretation of these results is similar to the one in Section 2.1. Namely, applying to equation (2.5.20) the operator $\nabla^2 + k^2 n_0^2(x)$ one gets

$$[\nabla^2 + k^2 n_0^2(x) - cN(x)h(x)]u = 0 \quad \text{in } \mathbb{R}^3. \qquad (2.5.21)$$

Thus,

$$n(x) = [n_0^2(x) - cN(x)h(x)k^{-2}]^{1/2} \qquad (2.5.22)$$

This formula is analogous to (2.1.32).

2.6   Summary of the results

Rather than repeating the formulation of theorems from Sections 2.1–2.5, let us discuss the essential points in a descriptive matter.

   In Chapter 2, we have given an analytical solution to the many-body wave scattering problems with various boundary conditions and proved that the effective field has a limit, u, as $a \to 0$, and this limit solves equation (2.1.27) in the case of the impedance boundary condition, and equation (2.2.11) in the case of the Dirichlet boundary condition. The limiting medium has a new refraction coefficient (2.1.32) in the case of the impedance boundary condition, and a refraction coefficient (2.1.13) in the case of the Dirichlet boundary condition.

In the case of the Neumann or the transmission boundary conditions, the limiting effective field does not satisfy a local differential equation.

The theory developed in this chapter allows one to solve many-body wave scattering problems for any number of small scatterers. If the number of the scatterers is very large, say, $M > 10^5$, then solving the limiting integral equation, for example, equation (2.1.27) in the case of the impedance boundary condition, can be a practically efficient way to solve the many-body wave scattering problem.

In Chapter 3, we apply the results of Chapter 2 to a theory of creating materials with a desired refraction coefficient by embedding in a given material many small particles with prescribed boundary impedances.

Chapter 3

Creating Materials with a Desired Refraction Coefficient

3.1 Scalar wave scattering: Formula for the refraction coefficient

Suppose that in a bounded domain filled with a material with a known refraction coefficient $n_0(x)$ M small ($ka \ll 1$) bodies $D_m, 1 \leq m \leq M$, of an arbitrary shape are embedded according to the distribution law (2.1.11), and the boundary impedance of these small bodies is given in (2.1.12). If conditions (2.1.22), (2.1.23), and (2.5.5) hold, then the effective field, acting on the jth body, is given by formula (2.5.18), where Q_m are defined in formula (2.1.14) and the numbers $u_m = u_e(x_m), x_m \in D_m$, are calculated by solving linear algebraic system (2.5.19), in which $G_{jm} = G(x_j, x_m)$ and $G(x, y)$ solves equation (2.5.6) and satisfies the radiation condition (2.5.7).

As $a \to 0$, the effective field tends to the unique solution of the integral equation (2.5.20). The limiting medium has a new refraction coefficient defined by formula (2.5.22):

$$n(x) = [n_0^2(x) - k^{-2}ch(x)N(x)]^{1/2}. \qquad (3.1.1)$$

The choice of the constant c is in the hands of the experimentalist. This constant is defined by the relation $|S_m| = ca^2$, where S_m is the surface area of a small body D_m and it is assumed that all the small bodies are of the same shape and size. If these bodies are balls, then $c = 4\pi$. The choice of the functions $N(x) \geq 0$ and $h(x), \mathrm{Im}\,h(x) \leq 0$, is also in the hands of the experimentalist.

The square root (3.1.1) of a complex number $z := n_0^2(x) - k^{-2}ch(x)N(x)$ is defined by the following formula:

$$z^{1/2} = |z|^{1/2}e^{i\frac{\phi}{2}}, \quad \phi = \arg z, \quad \phi \in [0, 2\pi). \quad (3.1.2)$$

If $\mathrm{Im}\, n_0^2(x) > 0$ and $\mathrm{Im}\, h(x) \leq 0$, then formulas (3.1.1) and (3.1.2) give $\arg n(x) \in [0, \frac{\pi}{2})$. If $n_0^2(x) = 0$ and $\mathrm{Im}\, h = 0$, then formulas (3.1.1) and (3.1.2) allow one to get $n(x) \in [0, n_0(x))$. Creating materials with a small refraction coefficient $n(x)$ is of practical interest: it may help to transform curved wave fronts into planar ones and has other possible applications.

By using formula (3.1.1), materials with such a refraction coefficient can be created by embedding into a given material many small impedance particles with prescribed boundary impedances. If the bodies are balls of radius a, then $c = 4\pi$. One can choose $N(x) = \text{const.}$, for example, $N(x) = 1$, and then choose

$$h(x) = \frac{n_0^2(x)}{4\pi k^{-2}} \quad (3.1.3)$$

to get material with $n(x) = 0$.

In the following section, we formulate a recipe for creating materials with a desired refraction coefficient.

3.2   A recipe for creating materials with a desired refraction coefficient

Problem 1: Given a material with known $n_0^2(x)$ in a bounded domain Ω and $n_0^2(x) = 1$ in Ω', one wants to create in Ω a material with a desired $n^2(x)$.

Step 1: Given $n_0^2(x)$ and $n^2(x)$, calculate

$$p(x) := k^2[n_0^2(x) - n^2(x)]. \quad (3.2.1)$$

This is a trivial step.

Step 2: Given $p(x)$, calculate $h(x)$ and $N(x)$ from the equation, see equation (3.1.1):

$$p(x) = ch(x)N(x). \tag{3.2.2}$$

One can take the constant c in (3.2.2) to be $c = 4\pi$. In this case, the small bodies D_m are balls, and (3.2.2) reduces to

$$p(x) = 4\pi h(x)N(x). \tag{3.2.3}$$

There are infinitely many solutions h and N to this equation. For example, one can fix arbitrarily $N(x) > 0$ in Ω, $N(x) = 0$ in Ω', and find $h(x) = h_1(x) + ih_2(x)$ by the formulas

$$h_1(x) = \frac{p_1(x)}{4\pi N(x)}, \quad h_2(x) = \frac{p_2(x)}{4\pi N(x)}. \tag{3.2.4}$$

Here,

$$p_1(x) = \operatorname{Re}p(x), \quad p_2(x) = \operatorname{Im}p(x). \tag{3.2.5}$$

Note that $\operatorname{Im}n^2(x) \geq 0$ implies $\operatorname{Im}h(x) \leq 0$, so our assumption $\operatorname{Im}h(x) \leq 0$ is satisfied. For example, one may take $N(x) = \text{const} > 0$. Step 2 is also trivial.

Step 3: Given $N(x)$ and $h(x) = h_1(x) + ih_2(x)$, distribute $M = O\left(\frac{1}{a^{2-\kappa}}\right)$ small impedance balls of radius a in the bounded region Ω according to the distribution law (2.1.11), where $\kappa \in [0,1)$ is the number (parameter) that can be chosen by the experimentalist. The minimal distance d between neighboring particles satisfies condition (2.1.10). Note that condition $d \gg a$ is satisfied automatically for the distribution law (2.1.11). Indeed, if d is the minimal distance between neighboring particles, then there are at most $\frac{1}{d^3}$ particles in a cube with the unit side, and since Ω is a bounded domain, there are at most $O\left(\frac{1}{d^3}\right)$ of small particles in Ω. On the other hand, by the distribution law (2.1.11), one has $\mathcal{N}(\Omega) = O\left(\frac{1}{a^{2-\kappa}}\right)$. Thus, $O\left(\frac{1}{d^3}\right) = O\left(\frac{1}{a^{2-\kappa}}\right)$. Therefore,

$$d = O\left(a^{\frac{2-\kappa}{3}}\right). \tag{3.2.6}$$

Consequently, condition (2.1.10) is satisfied. Moreover,

$$\lim_{a \to 0} \frac{a}{d(a)} = 0. \tag{3.2.7}$$

In the following section, we discuss the practical problems one should solve for implementing the above recipe for creating materials with a desired refraction coefficient.

3.3   A discussion of the practical implementation of the recipe

As mentioned in Section 3.2, the first two steps of the recipe are trivial both theoretically and practically. The only step that should be discussed is Step 3.

There are two technological problems that must be solved, and if they are solved, then practical implementation of our recipe is straightforward.

Technological Problem 1: How does one embed a large number $M = O(\frac{1}{a^{2-\kappa}})$ of small impedance particles into a given material in a bounded domain Ω according to the distribution law (2.1.11)?

Technological Problem 2: How does one prepare many small particles with prescribed values of the boundary impedances $\zeta_m = \frac{h(x_m)}{a^\kappa}$?

Technological Problem 1 seems to be solvable currently by the stereolithography process and also by the process of growth of small particles of desired sizes at the desired points.

Technological Problem 2 should be solvable because of the following argument based on physical reasoning.

If $\zeta = 0$, then the impedance boundary condition becomes the Neumann boundary condition, that is, "acoustically hard" body satisfies this condition. Thus, such boundary conditions make physical sense and the bodies which satisfy these conditions exist in reality.

The other limiting case of the impedance boundary condition is the Dirichlet boundary condition corresponding mathematically to the case $\zeta \to \infty$. It corresponds physically to "acoustically soft" bodies, and such bodies exist in reality. The mathematical limiting process $\zeta \to \infty$ is discussed in Section 11.3 in detail.

Thus, the bodies satisfying the impedance boundary condition with any $\zeta, \mathrm{Im}\zeta \leq 0$, should exist.

The general properties of the impedance are restricted physically only by the causality principle, see a discussion of these properties in Landau and Lifshitz (1984).

It would be of interest in practice to solve Technological Problem 2.

3.4   Summary of the results

Let us formulate the result in the following theorem.

Theorem 3.4.1. *Given $n_0^2(x)$ and a bounded domain Ω, one can create in Ω a material with a desired refraction coefficient $n(x), Imn^2(x) \geq 0$, by embedding $M = O\left(\frac{1}{a^{2-\kappa}}\right)$ small impedance particles according to the distribution law (2.1.11). The refraction coefficient $n_a(x)$, corresponding to a finite a, approximates the desired refraction coefficient $n(x)$ in the sense*

$$\lim_{a \to 0} n_a(x) = n(x). \tag{3.4.1}$$

The functions $h(x)$ and $N(x)$ defining the distribution law (2.1.11) are found by Steps 3.1 and 3.2 of the Recipe formulated in Section 3.2.

Finally, let us discuss briefly the possibility to create material with negative refraction coefficient. By formulas (3.1.1) and (3.1.2), one gets $n(x) < 0$ if the argument of $n_0^2(x) - k^{-2}cN(x)h(x)$ is equal to 2π. Assume that $n_0^2(x) > 0$. We know that $k^{-2}cN(x) \geq 0$. Let us take $h(x) = |h(x)|e^{i\varphi}$, where $0 < \varphi$ is very small, that is, $Imh(x) \geq 0$ and $Imh(x)$ is very small. We have proved in Section 2.1 that equation (2.1.27) is uniquely solvable for sufficiently small $Imh \geq 0$. For such $h(x)$, one has $\arg(n_0^2(x) - k^{-2}cN(x)h(x))$ which is very close to 2π, and the square root (3.1.1) is negative, provided that $Re(n_0^2(x) - k^{-2}cN(x)h(x)) > 0$. This argument shows that it is possible to create materials with negative refraction coefficient $n(x)$ by embedding in a given material many small particles with properly chosen boundary impedances.

Chapter 4

Wave-Focusing Materials

4.1 What is a wave-focusing material?

Consider the scattering problem

$$[\nabla^2 + k^2 - q(x)]u = 0 \quad \text{in } \mathbb{R}^3. \tag{4.1.1}$$

The scattering solution to equation (4.1.1) is of the form

$$u(x, \alpha, k) = e^{ik\alpha \cdot x} + A(\beta, \alpha, k)\frac{e^{ikr}}{r} + o\left(\frac{1}{r}\right), \quad r = |x| \to \infty, \quad \frac{x}{r} := \beta. \tag{4.1.2}$$

Here, $\alpha \in S^2$ is a given unit vector in the direction of propagation of the incident plane wave, the coefficient $A(\beta, \alpha, k) = A_q(\beta, \alpha, k)$ is called the scattering amplitude, β is the unit vector in the direction of the scattered wave. The function $A(\beta, \alpha, k)$ describes the angular distribution of the scattered field. The cross-section

$$\sigma(\alpha) = \int_{S^2} |A(\beta, \alpha, k)|^2 d\beta \tag{4.1.3}$$

describes the total energy scattered by the potential q when the incident field $e^{ik\alpha \cdot x}$ is incident upon the potential. If $q(x)$ is compactly supported, $q \in L^2$, and $\text{Im}q \leq 0$, then the scattering solution to equation (4.1.1) exists and is unique. This is proved as in the proof of Theorem 2.5.1.

From the mid-forties of the twentieth century, physicists wanted to prove that the observable quantities determine a physical system, that is, determine its Hamiltonian. For non-relativistic quantum-mechanical scattering, the potential describes the Hamiltonian, and the scattering amplitude $A(\beta, \alpha, k)$ is "the observable". This raises the following basic question:

Given the scattering amplitude, can one uniquely determine the potential q in equation (4.1.1)?

This question is not quite clearly stated because it is not specified for what β, α, and k the scattering amplitude is known.

If $A(\beta, \alpha, k)$ is known for all β and α on the unit sphere and for all $k > 0$, then it is easy to prove that q is uniquely determined by these data (see, for example, Saito (1986)).

If the scattering amplitude is known at a fixed $k > 0$ (a fixed energy $k^2 > 0$) and q is a compactly supported real-valued potential, $q \in L^2(\mathbb{R}^3)$, then the uniqueness of finding such a q from the fixed-energy scattering data was announced in Ramm (1987) and proved in Ramm (1988b), see also Ramm (1988a, 1988c, 1989, 1992, 1994, 2005a).

An algorithm for reconstruction of q from fixed-energy scattering data was presented by the author in Ramm (2002, 2005a). For many decades the uniqueness of the solution to inverse scattering problem of finding the potential from non-overdetermined scattering data was open. By non-overdetermined scattering data, one understands the value of the scattering set of the parameters β, α, k, for example, back-scattering data $A(-\alpha, \alpha, k), \alpha \in S^2, \forall k > 0$, or the fixed incident direction data $A(\beta, \alpha_0, k), \forall \beta \in S^2, \forall k > 0$, and a fixed $\alpha = \alpha_0 \in S^2$. Note that the unknown potential q depends on these variables.

The uniqueness of the solution to the inverse scattering problem with non-overdetermined data was first announced in Ramm (2009c) and proved in Ramm (2010d, 2010e, 2011c).

The inverse scattering problem with underdetermined data was not studied in the literature. By underdetermined scattering data, we understand the values of $A(\beta, \alpha, k)$ on a set of dimensions less than three. For example, consider the scattering data

$$A_q(\beta) := A(\beta) := A(\beta, \alpha_0, k_0), \quad \forall \beta \in S^2, \qquad (4.1.4)$$

that is, the values of the scattering amplitude at a fixed $k = k_0 > 0$, a fixed $\alpha = \alpha_0 \in S^2$, and all $\beta \in S^2$.

Inverse scattering problem (IP): *Given an arbitrary function* $f(\beta) \in L^2(S^2)$ *and an arbitrary small number* $\epsilon > 0$, *can one find a potential* $q \in L^2(\Omega)$ *such that*

$$||A_q(\beta) - f(\beta)|| \leq \epsilon, \qquad (4.1.5)$$

where the norm $|| \cdot ||$ *is* $L^2(S^2)$ *norm?*

We give a positive answer to this question and construct a potential $q \in L^2(\Omega)$ that generates the scattering data $A_q(\beta)$ satisfying (4.1.5).

A priori it is not clear that such a potential does exist. By Ω we mean an arbitrary but fixed bounded domain outside of which $q = 0$.

Note that the inverse scattering problem with underdetermined scattering data in general has many solutions. Therefore, we look for a solution. Since such problems were not studied earlier, there are open questions in the theory. Some of these questions will be formulated in this chapter.

In engineering literature the scattering amplitude $A(\beta, \alpha, k)$ is often called the radiation pattern.

A material with a wave-focusing property is a material which generates the scattering amplitude $A_q(\beta)$ satisfying inequality (4.1.5) in which $f(\beta)$ vanishes outside of a prescribed small solid angle.

The relation between the potential $q(x)$ and the refraction coefficient of the material is given as

$$k^2 n^2(x) = k^2 - q(x), \quad n^2(x) = 1 - k^{-2}q(x). \qquad (4.1.6)$$

In other words,

$$q(x) = k^2(1 - n^2(x)). \qquad (4.1.7)$$

It will be convenient in this chapter to deal with potentials q rather than with the refraction coefficients n, although there is a simple relation between these functions.

4.2 Creating wave-focusing materials

Our first result is the following theorem. Throughout we assume that Ω is a bounded domain in $\mathbb{R}^3$.

Theorem 4.2.1. *Let $f(\beta) \in L^2(S^2)$ be an arbitrary fixed given function, and $\epsilon > 0$ be an arbitrary small number. Then there exists a $q \in L^2(\Omega)$ such that inequality (4.1.5) holds.*

Proof. The proof will require several lemmas.

Lemma 4.2.1. *Suppose that the set $\{A_q(\beta)\}$ is dense in $L^2(S^2)$ when q runs through $L^2(\Omega)$. Then the set $\{A_q(\beta)\}$ is dense in $L^2(S^2)$ when q runs through $C_0^\infty(\Omega)$.*

Proof of Lemma 4.2.1. The proof is based on the following formula, proved by the author in Ramm (2002), Ramm (2005a), p. 262. This result is of independent interest:

$$-4\pi[A_{q_1}(\beta, \alpha, k) - A_{q_2}(\beta, \alpha, k)]$$
$$= \int_\Omega [q_1(x) - q_2(x)]u_1(x, \alpha)u_2(x, -\beta)dx, \qquad (4.2.1)$$

where $u_j(x, \alpha)$ is the scattering solution corresponding to the potential $q_j, j = 1, 2$.

We give a proof for formula (4.2.1) in what follows.

Assuming that formula (4.2.1) is proved and also that the following inequality

$$\sup_{\alpha \in S^2} \sup_{x \in \mathbb{R}^3} |u_j(x, \alpha)| \le c, \qquad (4.2.2)$$

is established, let us prove Lemma 4.2.1.

A proof of inequality (4.2.2) is given as follows.

To prove Lemma 4.2.1, note that (4.2.1) implies

$$||A_{q_1}(\beta) - A_{q_2}(\beta)||_{L^2(S^2)} \le c||q_1 - q_2||_{L^2(\Omega)}. \qquad (4.2.3)$$

Therefore, small variations of q in $L^2(\Omega)$ norm lead to small variations of $A_q(\beta)$ in $L^2(S^2)$ norm.

Since the set $C_0^\infty(\Omega)$ is dense in $L^2(\Omega)$, it follows that if a function $f \in L^2(S^2)$ can be approximated with an arbitrary accuracy in $L^2(S^2)$ by a function $A_q(\beta)$ with some $q_1 \in L^2(\Omega)$, then there is a potential $q_2 \in C_0^\infty(\Omega)$ which approximates q_1 with an arbitrary accuracy in the norm of $L^2(\Omega)$. By inequality (4.2.3) the function $A_{q_2}(\beta)$

approximates with any desired accuracy the original function $f(\beta)$ in $L^2(S^2)$.

Lemma 4.2.1 is proved. $\qquad\qquad\qquad\qquad\qquad\qquad\square$

Lemma 4.2.2. *Formula* (4.2.1) *holds.*

Proof. In this proof, we assume $k > 0$ arbitrary and fixed, and denote by $G(x, y)$ the resolvent kernel of the operator $-\nabla + q - k^2$, that is

$$(-\nabla + q(x) - k^2)G(x, y) = \delta(x - y), \quad \text{in } \mathbb{R}^3, \qquad (4.2.4)$$

where G satisfies the radiation condition.

The uniquely solvable integral equation for G is

$$G(x, y) = g(x, y) - \int_\Omega g(x, z)q(z)G(z, y)dz, \quad g(x, y) = \frac{e^{ik|x-y|}}{4\pi|x - y|}. \qquad (4.2.5)$$

Note that

$$g(x, y) = \frac{e^{ik|x|}}{4\pi|x|}e^{-ik\beta\cdot y}\left[1 + O\left(\frac{1}{|x|}\right)\right], \quad |x| \to \infty, \quad \beta := \frac{x}{|x|}, \qquad (4.2.6)$$

where y is any vector in a bounded domain and $e^{-ik\beta\cdot y}$ is the scattering solution $u_0(y, -\beta, k)$ for $q = 0$. One can prove, using equation (4.2.5), that

$$G(x, y) = \frac{e^{ik|x|}}{4\pi|x|}u(y, -\beta, k)\left(1 + O\left(\frac{1}{|x|}\right)\right), \quad |x| \to \infty, \quad \beta := \frac{x}{|x|}, \qquad (4.2.7)$$

where $u(y, -\beta, k)$ is the scattering solution, that is

$$\left(\nabla^2 + k^2 - q(x)\right)u(x, -\beta, k) = 0 \quad \text{in } \mathbb{R}^3, \qquad (4.2.8)$$

$$u(x, -\beta, k) = e^{-ik\beta\cdot x} + v, \qquad (4.2.9)$$

where v satisfies the radiation condition.

The existence of the representation (4.2.7) follows from equation (4.2.5) and formula (4.2.6). One can derive the following formula:

$$u(y, -\beta, k) = e^{-ik\beta \cdot y} - \int_\Omega e^{-ik\beta \cdot x} q(z) G(z, y) dz, \qquad (4.2.10)$$

from (4.2.5) and (4.2.6). The fact that function (4.2.10) is the scattering solution can be checked easily: the integral in (4.2.10) satisfies the radiation condition because $G(z, y)$ satisfies this condition, and the function (4.2.10) satisfies equation (4.2.8) because

$$(\nabla^2 + k^2) u(y, -\beta, k) = -\int_\Omega e^{-ik\beta \cdot z} q(z)[q(y)G(z, y) - \delta(z - y)]dz$$

$$= q(y)[u(y, -\beta, k) - e^{-ik\beta \cdot y}] + e^{-ik\beta \cdot y} q(y)$$

$$= q(y) u(y, -\beta, k). \qquad (4.2.11)$$

Since problems (4.2.8) and (4.2.9) have a unique solution, it follows that function (4.2.10) is the scattering solution.

To prove formula (4.2.1), we start with the formula

$$G_{q_2}(x, y) = G_{q_1}(x, y) - \int_\Omega G_{q_1}(x, z)[q_2(z) - q_1(z)]G_{q_2}(z, y)dz, \qquad (4.2.12)$$

that is verified easily by applying the operator $\nabla_x^2 + k^2 - q_1(x)$ to both sides of the formula (4.2.12). Let us take $|x| \to \infty$, $\frac{x}{|x|} = \beta$, and use formula (4.2.7) to get

$$u_2(y, -\beta, k) = u_1(y, -\beta, k) - \int_\Omega u_1(z, -\beta, k)[q_2(z)$$

$$- q_1(z)]G_{q_2}(z, y)dz. \qquad (4.2.13)$$

Now, let us take $|y| \to \infty$, $\frac{y}{|y|} = -\alpha$, and use formula (4.2.7) and formula (4.1.2) to get the following:

$$A_2(-\alpha, -\beta, k) = A_1(-\alpha, -\beta, k)$$

$$- \frac{1}{4\pi} \int [q_2(z) - q_1(z)] u_1(z, -\beta, k) u_2(z, \alpha, k) dz. \qquad (4.2.14)$$

Finally, take into account the known relation, the reciprocity property (see Ramm (1986)):

$$A(-\alpha, -\beta, k) = A(\beta, \alpha, k), \qquad (4.2.15)$$

and get from (4.2.14) the desired formula (4.2.1).

Lemma 4.2.2 is proved. □

Lemma 4.2.3. *Formula* (4.2.2) *holds.*

Proof of Lemma 4.2.3. The scattering solution solves the equation

$$u(x, \alpha, k) = e^{ik\alpha \cdot x} - \int_\Omega g(x, y)q(y)u(y, \alpha, k)dy := u_0 - Tu, \quad (4.2.16)$$

where the operator $T : C(\Omega) \to C(\Omega)$ and $(I + T)^{-1}$ is bounded. Therefore,

$$\sup_{x \in \Omega} |u(x, \alpha, k)| \le c \sup_{x \in \Omega} |u_0(x, \alpha, k)|, \qquad (4.2.17)$$

where the constant c does not depend on $\alpha \in S^2$.

Consequently, estimate (4.2.2) follows.

Lemma 4.2.3 is proved. □

This complete the proof of Lemma 4.2.1.

Remark 4.2.1. The results of Lemmas 4.2.1–4.2.3 belong to the author (see Ramm (1986) and Ramm (2005a)).

Let us continue with the proof of Theorem 4.2.1.

Lemma 4.2.4. *The set* $\{A_q(\beta)\}$ *is dense in* $L^2(S^2)$ *when* q *runs through* $C_0^\infty(\Omega)$.

Proof of Lemma 4.2.4. Suppose the conclusion of Lemma 4.2.4 is false and derive a contradiction from this assumption.

If the conclusion of Lemma 4.2.4 is false, then there exists a function $f(\beta) \in L^2(S^2), f \not\equiv 0$, such that

$$\int_{S^2} f(\beta)A_q(\beta)d\beta = 0, \quad \forall q \in C_0^\infty(\Omega). \qquad (4.2.18)$$

Recall that

$$A_q(\beta) = -\frac{1}{4\pi} \int_\Omega e^{-ik\beta \cdot y} q(y) u(y, \alpha, k) dy. \qquad (4.2.19)$$

From (4.2.18), it follows that

$$\int_\Omega dy q(y) u(y, \alpha, k) \int_{S^2} f(\beta) e^{-ik\beta \cdot y} d\beta = 0, \quad \forall q \in C_0^\infty(\Omega). \quad (4.2.20)$$

It follows from (4.2.20) and from the sufficient arbitrariness of q that

$$\int_{S^2} f(\beta) e^{-ik\beta \cdot y} d\beta = 0, \quad \forall y \in \Omega. \qquad (4.2.21)$$

Here we have used the fact that the scattering solutions do not vanish on an open subset of $\mathbb{R}^3$ as follows from the unique continuation property of the solutions to homogeneous Schrödinger equations.

It follows from (4.2.21) that $f = 0$. Indeed, the left-hand side of (4.2.21) is Fourier transform of a compactly supported distribution supported on the sphere of radius $k > 0$ with the density $f(\beta)$. This Fourier transform is an entire function of y which vanishes on an open subset $\Omega \subset \mathbb{R}^3$. Therefore, it vanishes identically in $\mathbb{R}^3$. By the injectivity of the Fourier transform, one concludes that $f(\beta) = 0$.

Lemma 4.2.4 is proved.

From Lemma 4.2.4, the conclusion of Theorem 4.2.1 follows. □

The question of interest now is to find a $q \in L^2(\Omega)$ such that inequality (4.1.5) holds. The existence of such a q follows from Theorem 4.2.1. Finding such a q amounts to solving the underdetermined inverse scattering problem that we have discussed in Section 4.1.

If $f(\beta) \in L^2(S^2)$ is an arbitrary given function, then one finds some $h(y) \in L^2(D)$ such that

$$\left\| f(\beta) + \frac{1}{4\pi} \int_\Omega h(y) e^{-ik\beta \cdot y} dy \right\|_{L^2(S^2)} \leq \epsilon. \qquad (4.2.22)$$

Here, $h(y) := q(y) u(y, \alpha, k)$, see formula (4.2.19). There are many $h \in L^2(\Omega)$ for which (4.2.22) holds. Indeed, if inequality (4.2.22) holds for some h, then it holds with 2ϵ in place of ϵ for any h_δ such that $\|h_\delta - h\|_{L^2(\Omega)} \leq \delta$, if $\delta > 0$ is sufficiently small.

We will discuss methods for finding $h \in L^2(\Omega)$ satisfying inequality (4.2.22) in Section 4.3.

If h is found from (4.2.22), then one would like to find a $q \in L^2(\Omega)$ such that

$$\|h(x) - qu(x, \alpha, k)\|_{L^2(\Omega)} \leq \delta, \qquad (4.2.23)$$

where $\delta > 0$ is an arbitrary small number, and $\alpha \in S^2$ is fixed. Recall that $k > 0$ is fixed also.

It is possible to satisfy inequality (4.2.23). This is proved in the following theorem.

Theorem 4.2.2. *The set* $\{q(x)u_q(x, \alpha, k)\}_{\forall q \in L^2(\Omega)}$ *is dense in* $L^2(\Omega)$ *and* $u_q(x, \alpha, k) := u(x, \alpha, k)$ *is the scattering solution corresponding to* q, *where* q *is a real-valued compactly supported potential.*

Proof. We assume that $\alpha \in S^2$ is fixed, $k > 0$ is fixed, and denote

$$q(x)u_q(x, \alpha, k) := h(x). \qquad (4.2.24)$$

We want to prove that the set $\{h(x)\}_{\forall q \in L^2(\Omega)}$ is dense in $L^2(\Omega)$. Note that if $u_0(x) = e^{ik\alpha \cdot x}$ and h is of the form (4.2.24), then

$$q(x) = \frac{h(x)}{u_q(x, \alpha, k)} = \frac{h(x)}{u_0(x) - \int_\Omega g(x, y)h(y)dy}. \qquad (4.2.25)$$

If $h \in L^2(\Omega)$ is arbitrary, then formula (4.2.25) defines some function $q(x)$ which may not belong to $L^2(\Omega)$. Since the denominator in (4.2.25) is in $H^2_{loc}(\mathbb{R}^3)$ if $h \in L^2(\Omega)$, the function $q(x)$ defined by formula (4.2.25) fails to be in $L^2(\Omega)$ if and only if the denominator vanishes at some set in $\mathbb{R}^3$. By $H^2_{loc}(\mathbb{R}^3)$, the Sobolev space of functions that belong to L^2 together with their derivatives up to the order two on every compact subset of $\mathbb{R}^3$.

If the denominator $u = u_q$ in (4.2.25) vanishes at some set of points in Ω, then

$$\text{Re}u(x) = 0, \quad \text{Im}u(x) = 0. \qquad (4.2.26)$$

These are two equations for twice differentiable functions $\text{Re}u$ and $\text{Im}u$ of $x = (x_1, x_2, x_3)$. These two equations define a line l in $\mathbb{R}^3$. Thus, the set of zeros of u is the set of measure zero in $\mathbb{R}^3$. Generally, the line l is smooth. Indeed, since we want to prove that the set of h of the form (4.2.25) is dense in $L^2(\Omega)$, one may assume $h \in C_0^\infty(\Omega)$, and in this case the functions $\mathrm{Re}u(x)$ and $\mathrm{Im}u(x)$ are $C^\infty(\Omega)$ functions, and l is smooth. The vectors $\nabla\mathrm{Re}u$ and $\nabla\mathrm{Im}u$ can be considered linearly independent and in this case the line l, defined by the equation (4.2.26), is smooth by the implicit function theorem. Recall that $\nabla\mathrm{Re}u$ and $\nabla\mathrm{Im}u$ are vectors directed along normals to the surfaces (4.2.26). We want to cover the line l, that is, the set of zeros of u, by a tubular neighborhood

$$N_\delta := \{x : x \in \Omega, |u(x)| \le \delta\}, \quad \Omega_\delta := \Omega \setminus N_\delta. \qquad (4.2.27)$$

Choose

$$h_\delta = \begin{cases} h & \text{in } \Omega_\delta, \\ 0 & \text{in } N_\delta. \end{cases} \qquad (4.2.28)$$

Then

$$
\begin{aligned}
q_\delta &= \frac{h_\delta(x)}{u_0(x) - \int_\Omega g(x,y)h_\delta(y)dy} \\
&= \begin{cases} \dfrac{h_\delta(x)}{u_0(x) - \int_{\Omega_\delta} g(x,y)h_\delta(y)dy} & \text{in } \Omega_\delta, \\ 0 & \text{in } N_\delta. \end{cases}
\end{aligned} \qquad (4.2.29)
$$

Since the measure of N_δ is arbitrarily small if δ is sufficiently small, one has

$$\|h - h_\delta\|_{L^2(\Omega)} = o(1), \quad \delta \to 0. \qquad (4.2.30)$$

Therefore, if (4.2.22) holds for h, then it will hold for h_δ with 2ϵ replacing ϵ, provided that $\delta > 0$ is sufficiently small.

Let us check that q_δ, defined in (4.2.29), belongs to $L^\infty(\Omega)$. One has

$$
\begin{aligned}
|q_\delta(x)| &\le \frac{\max_{x \in \Omega} |h_\delta(x)|}{\min_{x \in \Omega_\delta} |u_0(x) - \int_\Omega g(x,y)h_\delta(y)dy|} \\
&\le c \left(\min_{x \in \Omega_\delta} \left|u_0(x) - \int_\Omega g(x,y)h(y)dy\right| - \right. \\
&\qquad \left. \max_{x \in \Omega_\delta} \left|u_0(x) - \int_\Omega g(x,y)|h(y) - h_\delta(y)|dy\right| \right)^{-1}. \qquad (4.2.31)
\end{aligned}
$$

By definition (4.2.27), one has

$$\min_{x \in \Omega_\delta} \left| u_0(x) - \int_\Omega g(x,y) h(y) dy \right| = \min_{x \in \Omega_\delta} |u(x)| > \delta. \qquad (4.2.32)$$

One has

$$\int_\Omega g(x,y) |h(y) - h_\delta(y)| dy = \int_{N_\delta} \frac{|h(y) - h_\delta(y)|}{4\pi |x - y|} dy \le c \int_{N_\delta} \frac{dy}{|x - y|}, \qquad (4.2.33)$$

where

$$c = \frac{\max_{y \in \Omega} |h(y)|}{4\pi}. \qquad (4.2.34)$$

Let us prove that

$$I_\delta := \int_{N_\delta} \frac{dy}{|x - y|} \le c \delta^2 |\ln \delta|, \quad x \in \Omega_\delta, \quad \delta \to 0, \qquad (4.2.35)$$

where $c > 0$ is a constant. By c, we denote various constants independent of δ.

If (4.2.35) is proved, then

$$|q_\delta(x)| \le \frac{c}{\delta - c\delta^2 |\ln \delta|} = O\left(\frac{1}{\delta}\right), \qquad (4.2.36)$$

if δ is sufficiently small.

To prove (4.2.35), let us use the coordinates s_1, s_2, s_3, such that s_1 is directed along the normal to the surface $u_1 := \mathrm{Re}\, u = 0$, s_2 is directed along the normal to the surface $u_2 := \mathrm{Im}\, u = 0$, s_3 is directed along the tangent vector to the curve l, the origin of the coordinate system is located in l, and the s_1, s_2 axes are orthogonal to s_3 axis, that is s_3 axis is orthogonal to the plane in which axes s_1 and s_2 lie.

The Jacobian J of the transformation $(x_1, x_2, x_3) \to (s_1, s_2, s_3)$ at the origin is

$$J = \frac{\partial(u_1, u_2, x_3)}{(x_1, x_2, x_3)} \bigg|_{x=0} = \begin{vmatrix} u_{1,1} & u_{1,2} & 0 \\ u_{2,1} & u_{2,2} & 0 \\ 0 & 0 & 1 \end{vmatrix} \neq 0, \qquad (4.2.37)$$

where $u_{i,j} := \frac{\partial u_i}{\partial x_j}$, and we took into account that

$$\begin{vmatrix} u_{1,1} & u_{1,2} \\ u_{2,1} & u_{2,2} \end{vmatrix} \neq 0,$$

because the normals to the surfaces $u_1 = 0$ and $u_2 = 0$ are linearly independent, that is, ∇u_1 and ∇u_2 are linearly independent and are orthogonal to the line $x_3 = s_3$ at the origin.

In a neighborhood of the origin, one has in N_δ the inequality

$$|J| + |J^{-1}| \leq c. \tag{4.2.38}$$

The integral (4.2.35) can be written in the coordinates s_1, s_2, s_3, and estimated as follows:

$$I_\delta \leq c \int_0^\delta d\rho \int_0^{c_3} \frac{ds_3}{\sqrt{\rho^2 + s_3^2}}, \tag{4.2.39}$$

because the domain of integration can be described by the inequalities

$$u_1^2 + u_2^2 \leq \delta^2, \quad 0 \leq s_3 \leq c_3, \tag{4.2.40}$$

where $c_3 > 0$ is some constant. We have also used the inequalities

$$c_1 \left(u_1^2 + u_2^2 + s_3^2 \right) \leq |y|^2 \leq c_2 \left(u_1^2 + u_2^2 + s_3^2 \right), \tag{4.2.41}$$

which hold because of (4.2.38).

One has

$$\int_0^{c_3} \frac{ds_3}{\sqrt{\rho^2 + s_3^2}} = \ln \left(s_3 + \sqrt{\rho^2 + s_3^2} \right) \bigg|_0^{c_3} \leq c_4 + \ln \frac{1}{\rho} \leq c_5 \ln \frac{1}{\rho}. \tag{4.2.42}$$

Therefore, estimate (4.2.35) follows from (4.2.39) and (4.2.42). Consequently, inequality (4.2.31) yields

$$|q_\delta(x)| \leq \frac{c}{\delta - O(\delta^2 |\ln \delta|)} = O \left(\frac{1}{\delta} \right). \tag{4.2.43}$$

Therefore, if formula (4.2.25) does not give $q \in L^2(\Omega)$, then a small perturbation of h, defined in (4.2.28), yields a function q_δ bounded in Ω.

We have proved that the set $\{h_\delta\} = \{q_\delta u_\delta\}$ is dense in $L^2(D)$, where q_δ are bounded functions.

This proves Theorem 4.2.2. $\qquad\qquad\qquad\qquad\qquad\qquad\qquad$ □

Remark 4.2.2. Let $H_0^2(\Omega)$ be the Sobolev space of twice differentiable functions whose second derivatives belong to $L^2(\Omega)$ and the functions vanish in a neighborhood of the boundary $\partial\Omega$ of Ω. Then

$$\int_\Omega e^{-ik\beta\cdot x}(\nabla^2 + k^2)\Phi dx = 0 \quad \forall \Phi \in H_0^2(\Omega), \qquad (4.2.44)$$

as one can check by using Green's formula and taking into account that

$$\Phi = \frac{\partial\Phi}{\partial N} = 0 \quad \text{on } \partial\Omega. \qquad (4.2.45)$$

Therefore,

$$\int_\Omega g(x,y)[h(y) + (\nabla^2 + k^2)\Phi]dy = \int_\Omega g(x,y)h(y)dy - \Phi(x),$$
$$(4.2.46)$$

where we have use Green's formula, took into account the equation $(\nabla^2 + k^2)g = -\delta(x-y)$, and the boundary conditions (4.2.45).

Consequently, the potentials $q(x)$ and $q_\Phi(x)$ generate the same scattering amplitude.

Here $\Phi \in H_0^2(\Omega)$ and

$$q_\Phi(x) = \frac{h(y) + (\nabla^2 + k^2)\Phi}{u_0 - \int_\Omega g(x,y)h(y)dy + \Phi(x)},$$

$$q(x) = \frac{h(y)}{u_0 - \int_\Omega g(x,y)h(y)dy}. \qquad (4.2.47)$$

One may try to use this property in order to find Φ such that $\mathrm{Im}q_\Phi \leq 0$. This is of interest because the inequality $\mathrm{Im}q_\Phi \leq 0$ guarantees the uniqueness of the scattering solutions. The problem of the existence of such Φ is currently open.

4.3 Computational aspects of the problem

There are two problems that we discuss from the computational point of view.

The first problem consists of finding $h(x) \in \Omega$ such that

$$\left\| f(\beta) + \frac{1}{4\pi} \int_\Omega e^{-ik\beta \cdot y} h(y) dy \right\|_{L^2(S^2)} \le \epsilon. \qquad (4.3.1)$$

Here, $f \in L^2(S^2)$ and a small number $\epsilon > 0$ are given, and $h \in L^2(\Omega)$ is to be found. Existence of such h is proved in Theorem 4.2.1. There are many h for which inequality (4.3.1) holds. Let us give an algorithm for finding such an h. Let us assume without loss of generality that Ω is a ball $B_R := B$ centered at the origin and of radius R. This can be assumed without loss of generality because one can always choose $B \subset \Omega$, and if $h \in L^2(B)$ is found, which satisfies inequality (4.3.1), then this h can be extended to $\Omega \setminus B$ by zero, it will be then defined in Ω, it will belong to $L^2(\Omega)$, and it will satisfies inequality (4.3.1).

Expand the plane wave $e^{-ik\beta \cdot y}$ and $h(y)$ into the following spherical harmonic series:

$$e^{-ik\beta \cdot y} = \sum_{l=0}^\infty 4\pi i^l j_l(kr) Y_l(-\beta)\overline{Y_l(y^0)}, \quad r := |y|, \ y^0 := \frac{y}{r}, \qquad (4.3.2)$$

$$j_l(r) := \left(\frac{\pi}{2r}\right)^{1/2} J_{l+\frac{1}{2}}(r),$$

$$\frac{4\pi}{2l+1} \sum_{m=-l}^l Y_{l,m}(x^0)\overline{Y_{l,m}(y^0)} = P_l(x^0 \cdot y^0). \qquad (4.3.3)$$

Here, $J_{l+\frac{1}{2}}(r)$ is the Bessel function regular at $r = 0, Y_l(\beta) = Y_{lm}(\beta), -l \le m \le l$, are the spherical harmonics, the over bar $\overline{Y_l(y^0)}$ stands for complex conjugate,

$$Y_{l,m}(-\beta) = (-1)^l Y_{l,m}(\beta), \quad \overline{Y_{l,m}(\beta)} = (-1)^{l+m} Y_{l,-m}(\beta), \qquad (4.3.4)$$

$$Y_{l,m}(\alpha) := \frac{(-1)^{m+|m|} i^l}{\sqrt{4\pi}} \left[\frac{(2l+1)(l-|m|)!}{(l+|m|)!} \right]^{1/2} e^{im\phi} P_{l,m}(\cos\theta), \qquad (4.3.5)$$

$$P_{l,m}(\cos\theta) := (\sin\theta)^m \frac{d^m P_l(\cos\theta)}{(d\cos\theta)^m}, \quad P_l(t) := \frac{1}{2^l l!} \frac{d^l (t^2-1)^l}{dt^l}, \qquad (4.3.6)$$

where $t = \cos\theta$, $P_l(t)$ are Legendre polynomials, $P_{l,m}(\cos\theta)$ are associated Legendre functions (see Lebedev (1972)), the unit vector β is described by the spherical coordinates (θ, ϕ), $0 \le \theta \le \pi$, $0 \le \phi \le 2\pi$, the summation in (4.3.2) and below $\sum_{l=0}^{\infty}$ is understood as $\sum_{l=0}^{\infty} \sum_{m=-l}^{l}$, and

$$(Y_{l,m}, Y_{l',m'})_{L^2(S^2)} = \delta_{ll'}\delta_{mm'}. \tag{4.3.7}$$

Let

$$h(y) = \sum_{l=0}^{\infty} c_l(r)Y_l(y^0), \quad r = |y|, \quad y^0 = \frac{y}{r}, \tag{4.3.8}$$

where the Fourier coefficients $c_l(r)$ are defined by the formula

$$c_l(r) := \int_{S^2} h(y)\overline{Y_l(y^0)}dy^0. \tag{4.3.9}$$

Let

$$f_L(\beta) = \sum_{l=0}^{L} f_l Y_l(\beta), \tag{4.3.10}$$

where L is a sufficiently large integer such that

$$\|f - f_L\|_{L^2(S^2)} < \frac{\epsilon}{2}. \tag{4.3.11}$$

Thus, if

$$\left\| f_L(\beta) + \frac{1}{4\pi} \int_B e^{-ik\beta \cdot y} h(y)dy \right\| < \frac{\epsilon}{2}, \tag{4.3.12}$$

then inequality (4.3.1) holds.

Let us find an h satisfying inequality (4.3.12). Consider the equation for h_l as follows:

$$\int_B e^{-ik\beta \cdot y} h_L(y)dy = -4\pi f_L(\beta). \tag{4.3.13}$$

Substitute (4.3.8) and (4.3.10) into (4.3.13) and use formulas (4.3.2), (4.3.4) and (4.3.7) to get

$$(-i)^{l+2} \int_0^R r^2 j_l(kr) c_l(r) dr = f_l, \qquad (4.3.14)$$

where $c_l(r) = c_{l,m}(r)$, $f_l = f_{l,m}$. Equation (4.3.14) has many solutions. Let $c_l^{\perp}(r)$ be any function orthogonal to $r^2 j_l(kr)$, that is,

$$\int_0^R r^2 j_l(kr) c_l^{\perp}(r) dr = 0. \qquad (4.3.15)$$

The general solution to equation (4.3.14) is

$$c_l(r) = a_l f_l j_l(kr) + b_l c_l^{\perp}(r), \qquad (4.3.16)$$

where b_l is an arbitrary constant and

$$a_l = \frac{1}{(-i)^{l+2} \int_0^R r^2 j_l^2(kr) dr}. \qquad (4.3.17)$$

We have proved the following theorem:

Theorem 4.3.1. *If L is chosen so that (4.3.11) holds, and $c_l(r)$ are defined by the formula*

$$c_l(r) = a_l f_l j_l(kr), \qquad (4.3.18)$$

where a_l is defined in (4.3.17), then the function

$$h_L(y) := \sum_{l=0}^L c_l(r) Y_l(y^0), \quad r = |y|, \quad y^0 := \frac{y}{r}, \qquad (4.3.19)$$

satisfies the inequality (4.3.1).

Let us give a formula for a potential q such that qu_q approximates h_L, defined in (4.3.19), with a desired accuracy. This h_L we denote by h in what follows. Using the h, defined in (4.3.19), calculate the potential q by formula (4.2.25). If this q belongs to $L^2(B)$, then the underdetermined inverse problem (IP) is solved.

If $q \notin L^2(B)$, then compute h_δ in place of h by formula (4.2.28) and q_δ by formula (4.2.29). This q_δ is bounded, and the corresponding to q_δ scattering amplitude $\frac{1}{4\pi} \int_B e^{-ik\beta \cdot y} h_\delta(y) dy$ satisfies inequality (4.3.1).

One can suggest a different computational method for finding $h(y)$ if $f(\beta)$ is given. For example, choose a basis $\{\phi_j\}$ in $L^2(B)$ and look for h_n, as follows:

$$h_n(x) = \sum_{j=1}^{n} a_j^{(n)} \phi_j(x), \qquad (4.3.20)$$

where $a_j^{(n)}$ are some coefficients which are found from the following minimization problem:

$$\left\| f(\beta) - \sum_{j=1}^{n} a_j^{(n)} g_j(\beta) \right\|_{L^2(S^2)} = \min, \qquad (4.3.21)$$

where

$$g_j := g_j(\beta) := -\frac{1}{4\pi} \int_B e^{-ik\beta \cdot y} \phi_j(y) dy. \qquad (4.3.22)$$

A necessary condition for the coefficients $a_j^{(n)}$ to yield minimum in (4.3.21) is a linear algebraic system

$$\sum_{j=1}^{n} (g_j, g_m) a_j^{(n)} = (f, g_m), \quad 1 \le m \le n. \qquad (4.3.23)$$

We assume that the set $\{g_j(\beta)\}$, where $g_j(\beta)$ are defined in (4.3.22), is a linearly independent set. This is a nontrivial assumption since the equation

$$\int_B e^{-ik\beta \cdot y} \phi(y) dy = 0, \quad \forall \beta \in S^2, \qquad (4.3.24)$$

where $k = \text{const} > 0$ is fixed, has many nontrivial solutions. Therefore, the suggested method, which seems simple, requires checking that the set $\{g_j\}_{j=1}^{n}$ is linearly independent.

4.4   Open problems

Our construction of q given the underdetermined scattering data did not resolve the following problems, which are currently open:

1. Can our algorithm for finding q yield a real-valued q?
2. Can it yield a q such that $\text{Im} q \leq 0$?

Both questions are of interest because we have proved uniqueness of the scattering solutions assuming $\text{Im} q \leq 0$.

4.5   Summary of the results

In this chapter, the following underdetermined inverse problem (IP) is investigated and solved:

(IP): *Given an arbitrary* $f \in L^2(S^2)$ *and an arbitrary small fixed* $\epsilon > 0$, *can one find* $q \in L^2(\Omega)$, *where* $\Omega \in \mathbb{R}^3$ *is an arbitrary fixed bounded domain, such that inequality* (4.1.5) *holds, where*

$$A_q(\beta) = -\frac{1}{4\pi} \int_\Omega e^{-ik\beta \cdot y} q(y) u(y, \alpha, k) dy, \qquad (4.5.1)$$

$k > 0$ *and* $\alpha \in S^2$ *are fixed?*

It is proved that the IP has a solution and there are infinitely many solutions to this IP. Some formulas for a solution to IP are given.

Chapter 5

Electromagnetic Wave Scattering by a Single Small Body of an Arbitrary Shape

5.1 The impedance boundary condition

Consider the following scattering problem for electromagnetic (EM) waves:

$$\nabla \times E = i\omega\mu H, \quad \text{in } D' := \mathbb{R}^3 \setminus D, \qquad (5.1.1)$$

$$\nabla \times H = -i\omega\epsilon' E, \quad \text{in } D', \qquad (5.1.2)$$

$$[N, [E, N]] = \zeta[N, H], \quad \text{Re}\zeta \geq 0, \qquad (5.1.3)$$

$$E = E_0 + v_E, \qquad (5.1.4)$$

$$E_0 = \mathcal{E}e^{ik\alpha \cdot x}, \quad \mathcal{E} \cdot \alpha = 0, \quad \alpha \in S^2, \qquad (5.1.5)$$

$$\frac{\partial v_E}{\partial r} - ikv_E = o\left(\frac{1}{r}\right), \quad r := |x| \to \infty. \qquad (5.1.6)$$

Condition (5.1.6) is called the radiation condition. It holds uniformly with respect to the directions $x^0 := \frac{x}{r}$. Condition (5.1.5) means that the incident field E_0 satisfies the condition $\nabla \cdot E_0 = 0$. Let us assume that the dielectric permittivity ϵ and magnetic permittivity μ are constants, D is a bounded domain with a sufficiently smooth boundary S, for example, S is of class $C^{1,\gamma}, \gamma > 0$ and $\epsilon' = \epsilon + \frac{i\sigma}{\omega}$.

Here, $\sigma \geq 0$ is a conductivity, $w > 0$ is a frequency, $k = \frac{2\pi}{\lambda}$, λ is the wave length, ζ is the boundary impedance , $C^{1,\gamma}$ is the class of surfaces whose equations in the local coordinates are of the form $z = f(x, y)$ with $f \in C^{1,\gamma}$. The local coordinates are the coordinates whose origins are at a point on the surface S, z-coordinate is directed along the normal to S at the origin and points out into D', and x,y-coordinates are in the plane tangent to S at the origin. The class $C^{1,\gamma}$ consists of the functions whose first derivatives satisfy the Hölder condition with the exponent $\gamma, 0 < \gamma \leq 1$, that is

$$|\nabla f(x, y) - \nabla f(x', y')| \leq c(|x - x'|^2 + |y - y'|^2)^{\gamma/2}. \qquad (5.1.7)$$

Our goal is to derive an explicit analytical formula for the solution to problems (5.1.1)–(5.1.6). No such formulas were known except for Mie-type solution for spheres, which was obtained by separation of variables. It is not possible to obtain such a solution without some assumptions. Our basic assumption is the assumption concerning the smallness of D. Let $a = \frac{1}{2}\mathrm{diam}D$. Then we assume that

$$ka \ll 1. \qquad (5.1.8)$$

No analytical results for EM wave scattering by small bodies of an arbitrary shape were obtained by other authors.

Let us first prove existence and uniqueness of the solution to problems (5.1.1)–(5.1.6). The smallness assumption (5.1.8) is not used in Theorem 5.1.1.

Theorem 5.1.1. *Problems* (5.1.1)–(5.1.6) *have no more than one solution.*

Proof. Since the problem is linear, it is sufficient to prove that the corresponding homogeneous problem, that is, the problem with $E_0 = 0$, has only the trivial solution.

Let us note first that the radiation condition (5.1.6) holds also for H, and implies that

$$[e_r, E] = \sqrt{\frac{\mu}{\epsilon}}H + o\left(\frac{1}{|x|}\right), \quad [H, e_r] = \sqrt{\frac{\epsilon}{\mu}}E + o\left(\frac{1}{|x|}\right),$$

$$r = |x| \to \infty. \qquad (5.1.9)$$

Here, e_r is the unit vector along the radius vector, e_r is the unit normal to the sphere $S_r = \{x : |x| = r\}$.

Multiply equation (5.1.1) by $\overline{H}$, complex conjugate of equation (5.1.2) by E, subtract from the first equation the second, and integrate over the domain $D_R := B_R \setminus D$. The result is

$$\int_{D_R} (\overline{H} \cdot \nabla \times E - E \cdot \nabla \times \overline{H}) dx = i\omega \int_{D_R} (\mu |H|^2 - \overline{\epsilon'} |E|^2) dx.$$

(5.1.10)

Note that the integrand on the left of (5.1.10) is $\nabla \cdot [E, \overline{H}]$, and apply the divergence theorem to get

$$\int_{S_r} N \cdot [E, \overline{H}] ds - \int_S N \cdot [E, \overline{H}] ds = i\omega \int_{D_R} (\mu |H|^2 - \overline{\epsilon'} |E|^2) dx.$$

(5.1.11)

Using formulas (5.1.9) and (5.1.3), transform the left-hand side of (5.1.11) and get

$$\int_{S_r} |E|^2 ds + o(1) + \zeta \int_S (|H|^2 - |N \cdot H|^2) ds$$
$$= i\omega \int_{D_r} \left(\mu |H|^2 - \epsilon |E|^2 + \frac{i\sigma}{\omega} |E|^2 \right) dx.$$

(5.1.12)

Here, we have used the formula $[N, E] = \zeta[N, [N, H]]$, which follows from the boundary condition (5.1.3), and get

$$-\int_S N \cdot [E, \overline{H}] ds = \zeta \int_S (|H|^2 - |N \cdot H|^2) ds.$$

(5.1.13)

Taking the real part of (5.1.12), one obtains

$$\lim_{r \to \infty} \int_{S_r} |E|^2 ds + \text{Re}\zeta \int_S (|H|^2 - |N \cdot H|^2) ds$$
$$= -\lim_{r \to \infty} \int_{D_r} \sigma(x) |E|^2 dx.$$

(5.1.14)

We assume that

$$\text{Re}\zeta \geq 0, \quad \sigma \geq 0.$$

(5.1.15)

Therefore, (5.1.14) is only possible if $E = 0$ in D' in the case $\sigma(x) > 0$, or

$$\lim_{r \to \infty} \int_{S_r} |E|^2 ds = 0, \qquad (5.1.16)$$

in the case $\text{Re}\zeta = 0$ and $\sigma = 0$. If (5.1.16) holds, then we derive that $E = 0$ in D'. Indeed, outside a large ball B_R, we have $\sigma = 0$, ϵ and μ are constants, so $\nabla \cdot E = \frac{\nabla \cdot \nabla \times H}{-i\omega\epsilon}$, and the equation

$$\nabla \times \nabla \times E = k^2 E, \quad k^2 = \omega^2 \epsilon \mu, \qquad (5.1.17)$$

which follows from (5.1.1) and (5.1.2) by applying curl to (5.1.1) and assuming $\mu = \text{const}$.

If $\nabla \cdot E = 0$, then equation (5.1.17) reduces to

$$(\nabla^2 + k^2)E = 0 \qquad (5.1.18)$$

because $\nabla \times \nabla \times E = \nabla\nabla \cdot E - \nabla^2 E = -\nabla^2 E$. Equation $\nabla \cdot E = 0$ holds if $\epsilon' = \text{const}$, as follows from equation (5.1.2) by taking divergence.

Any solution to equation (5.1.18) satisfying condition (5.1.16) is equal to zero in the region where equation (5.1.18) holds (see Lemma 1 on p. 25 in Ramm (1986)).

Thus, $E = 0$ for $|x| > R$. If $\epsilon' = \text{const}$, then equation (5.1.17) reduces to the elliptic equation (5.1.18) in D' and one can apply the unique continuation principle, which says: if a solution to equation (5.1.18) vanishes on an open set, then it vanishes everywhere in the domain in which equation (5.1.18) holds. Thus, in our case, $E = 0$ in D'.

Theorem 5.1.1 is proved.   □

To prove that problems (5.1.1)–(5.1.6) have a solution, it is sufficient to show that this problem is of Fredholm type (with index zero) since we have already proved that the corresponding homogeneous problem has only the trivial solution.

Let us reduce our problem to a Fredholm-type problem assuming that μ and ϵ' do not vanish and neglecting the boundary condition on S. Take divergence of equations (5.1.1)–(5.1.2) and get

$$\nabla \cdot (\mu H) = 0, \quad \nabla \cdot (\epsilon' E) = 0. \qquad (5.1.19)$$

We do not assume here that μ and ϵ' are constants. Taking curl of equation (5.1.1) and using equation (5.1.2) yield

$$\nabla \times \nabla \times E = i\omega[\nabla\mu, H] + i\omega\mu\nabla \times H = K^2(x)E + \left[\frac{\nabla\mu}{\mu}, \nabla \times E\right],$$
$$(5.1.20)$$

where $K^2 := \omega^2\epsilon'\mu$. One has

$$\nabla \times \nabla \times E = \nabla\nabla \cdot E - \nabla^2 E = -\nabla^2 E + \nabla\left(\frac{\nabla\epsilon'}{\epsilon'} \cdot E\right), \quad (5.1.21)$$

where $\nabla \cdot E$ is found from equation (5.1.19)

$$\nabla \cdot E = -\frac{\nabla\epsilon'}{\epsilon'} \cdot E \qquad (5.1.22)$$

and we assume that ϵ' is twice continuously differentiable, μ is continuously differentiable, ϵ' and μ are constants outside an arbitrary large but finite ball B_R.

Thus, equation (5.1.20) can be written as a second-order elliptic equation

$$-\nabla^2 E - \nabla\left(\frac{\nabla\epsilon'}{\epsilon'} \cdot E\right) - \left[\frac{\nabla\mu}{\mu}, \nabla \times E\right] - K^2(x)E = 0. \quad (5.1.23)$$

This equation, the boundary condition (5.1.3), and the radiation condition (5.1.6) form a Fredholm-type problem in the space $L^2(D', w)$ where the weight $w = \frac{1}{1+|x|^{1+\delta}}$, where $\delta > 0$. We did not discuss in this argument the role of the boundary condition on S. This is done in Section 5.3.

5.2 Perfectly conducting bodies

If $\zeta = 0$ in the impedance boundary condition (5.1.3), then one has a perfectly conducting body. The boundary condition for such a body is

$$E^t = [N[E, N]] = 0 \quad \text{on } S, \qquad (5.2.1)$$

where E^t is the tangential to S component of E,

$$E^t := E - NN \cdot E = [N[E, N]] = 0 \quad \text{on } S. \qquad (5.2.2)$$

Theorem 5.1.1 remains valid for perfectly conducting bodies. The proofs are also particular cases of the proofs given in Section 5.1.

Let us derive the physical meaning of $Q = \int_S J(t)dt$ for perfect conductors. To derive this physical meaning, remember the definition of the polarization $P = \int_D x\rho(x)dx$, where $\rho = \rho(x)$ is the charge density. Consider

$$-i\omega P = \int_D x(-i\omega\rho)dx$$

and use the continuity equation

$$-i\omega\rho + \nabla \cdot (\rho\vec{v}) = 0.$$

Since $\rho\vec{v} = J$, where $\vec{v}$ is the charge velocity, one gets

$$-i\omega P = -\int_D x\nabla \cdot Jdx = \int_D Jdx - \int_S tN \cdot Jdt = \int_D Jdx = Q,$$

because $N \cdot J = 0$ on S. If J is concentrated on S, then one gets $\int_S J(t)dt = Q$.

The physical meaning of Q is clear from the above, more precisely, when the time dependence is given by the factor $e^{-i\omega t}$, as we assume, then $-i\omega P$ corresponds to $\frac{\partial P}{\partial t}$.

5.3 Formulas for the scattered field in the case of EM wave scattering by one impedance small body of an arbitrary shape

In this section, we assume that

$$ka \ll 1, \quad \mathrm{Re}\,\zeta \geq 0. \tag{5.3.1}$$

Our goal is to give an analytic solution to problems (5.1.1)–(5.1.6) for a small body of an arbitrary shape.

Let us look for the solution of this problem of the form

$$E(x) = E_0(x) + \nabla \times \int_S g(x,t)J(t)dt, \quad g(x,t) := \frac{e^{ik|x-t|}}{4\pi|x-t|},$$
$$\tag{5.3.2}$$

where E_0 is the incident plane wave defined in (5.1.5) and J is an unknown pseudovector. We assume J to be tangential to S and twice continuously differentiable. Recall that E is a vector, $\nabla \times E$ is a pseudovector. This means that the coordinate transformation $x \to -x$, an inverse (reflection with respect to the origin) while changing the sign of a vector keeps a pseudovector unchanged. Since equation (5.1.1) shows that H is proportional to $\nabla \times E$, it follows that H is a pseudovector.

Since E is a vector, the unknown J in equation (5.3.2) is a pseudovector.

The right-hand side of (5.3.2) for any J satisfies the equation (5.1.17), the conditions (5.1.5)–(5.1.6), and if J is found from the impedance boundary condition (5.1.3), then H is found by the formula

$$H = \frac{\nabla \times E}{i\omega\mu}, \tag{5.3.3}$$

and the pair $\{E, H\}$ solves problems (5.1.5)–(5.1.6). Theorem 5.1.1 guarantees the uniqueness of the solution. Formula (5.1.17) together with formula (5.3.3) yield formula (5.1.2) if one takes into account that $\frac{k^2}{i\omega\mu} = -i\omega\epsilon'$. Therefore, it is sufficient to deal only with finding E by formula (5.3.2), that is, with finding J from condition (5.1.3).

We need the following known formula (see Chapter 11):

$$\left[N_s, \nabla \times \int_S g(x,t) J(t) dt \right]_{\mp} = \int_S [N_s, [\nabla_s g(s,t), J(t)]] dt \pm \frac{J(s)}{2}, \tag{5.3.4}$$

where the sign $\mp$ denote limiting values of the left-hand side of (5.3.4) as $x \to s \in S$ from outside (inside) D along the normal N_s.

Let us substitute (5.3.2) into the boundary condition (5.1.3) and use (5.3.4) to get the following equation:

$$f + \left[N_s, \left[\nabla \times \int_S g(s,t) J(t) dt, N_s \right] \right]$$
$$- \frac{\zeta}{i\omega\mu} \left[N_s, \nabla \times \nabla \times \int_S g(s,t) J(t) dt \right] = 0. \tag{5.3.5}$$

Here, we substitute in (5.1.3) for H its expression (5.3.3) and define

$$f := [N, [E_0, N]] - \frac{\zeta}{i\omega\mu}[N, \nabla \times E_0]. \qquad (5.3.6)$$

Thus, f is known since E_0 is known and S is known.

We do not intend to solve equation (5.3.5) for J. Instead, our goal is to find one pseudovector

$$Q = \int_S J(t)dt. \qquad (5.3.7)$$

Before we find an analytic formula for Q, let us explain the significance of Q.

If E is found of the form (5.3.2), then

$$E(x) = E_0(x) + [\nabla_x g(x, x_1), Q], \quad a \to 0, \qquad (5.3.8)$$

where we have shown the main term of E as $a \to 0$. Formula (5.3.8) holds in the region $|x - x_1| \gg a$, that is, practically everywhere as $a \to 0$. The point $x_1 \in D$ is an arbitrary point inside the small body D.

The EM wave scattering problems (5.1.1)–(5.1.6) are reduced by formula (5.3.8), valid for small bodies, to finding an unknown pseudo vector Q rather than an unknown function $J(t)$ on S. This is a drastic simplification of the problem. Formula (5.3.8) is asymptotically exact, as we will prove. Then we derive an explicit analytical formula for Q.

Lemma 5.3.1. *Formula (5.3.8) is asymptotically exact as $a \to 0$.*

Proof.   Let us write

$$\mathcal{I} := \nabla \times \int_S g(x, t)J(t)dt = \nabla \times g(x, x_1) \int_S J(t)dt + \nabla$$

$$\times \int_S [g(x, t) - g(x, x_1)]J(t)dt, \qquad (5.3.9)$$

where $x_1 \in D, |x_1 - t| = O(a)$. One can rewrite it as

$$\mathcal{I} = [\nabla \times g(x, x_1), Q] + I_2 := I_1 + I_2. \qquad (5.3.10)$$

Let us prove that

$$|I_2| \ll |I_1|, \quad a \to 0. \tag{5.3.11}$$

One has $|\nabla g(x, x_1)| = O\left(\max\left(\frac{1}{d^2}, \frac{k}{d}\right)\right)$, where $d = |x - x_1| \gg a$, so

$$|I_1| = QO\left(\max\left(\frac{1}{d^2}, \frac{k}{d}\right)\right), \tag{5.3.12}$$

and

$$|I_2| = QO\left(\max\left(\frac{a}{d^3}, \frac{ak^2}{d}\right)\right). \tag{5.3.13}$$

Thus,

$$\frac{|I_2|}{|I_1|} \leq O\left(\max\left(\frac{a}{d}, ka\right)\right). \tag{5.3.14}$$

Thus, as $a \to 0$, one gets estimate (5.3.11).

Lemma 5.3.1 is proved. $\qquad\qquad\qquad\qquad\qquad\qquad\qquad\qquad\square$

Let us now derive the analytical formula for Q.

Theorem 5.3.1. *One has*

$$Q = -\frac{\zeta|S|}{i\omega\mu}\tau\nabla \times E_0, \tag{5.3.15}$$

where

$$\tau := I - b, \quad b = (b_{jm}) := \frac{1}{|S|}\int_S N_j(s)N_m(s)ds, \tag{5.3.16}$$

and $|S|$ is the surface area of S.

Remark 5.3.1. If one assumes that

$$\zeta = \frac{h}{a^\kappa}, \quad 0 \leq \kappa < 1, \quad \text{Re}h \geq 0, \tag{5.3.17}$$

then

$$Q = O(a^{2-\kappa}), \quad a \to 0. \tag{5.3.18}$$

The constants h and κ can be chosen by the experimenter.

Proof of Theorem 5.3.1. Take a vector product of N with equation (5.3.5) and integrate the result over S. We obtain

$$I_0 + I_1 + I_2 = 0, \quad I_0 := \int_S [N, f] ds, \tag{5.3.19}$$

$$I_1 := \int_S \left[N, \left[N, \left[\nabla \times \int_S gJdt, N \right] \right] \right] ds = \int_S \left[N, \nabla \times \int_S gJdt \right] ds, \tag{5.3.20}$$

$$I_2 := -\frac{\zeta}{i\omega\mu} \int_S \left[N_s, \left[N_s, \nabla \times \nabla \times \int_S g(s,t)J(t)dt \right] \right] ds. \tag{5.3.21}$$

Let us estimate each of these three integrals and keep the main terms as $a \to 0$.

One has

$$I_0 = I_{00} + I_{01}, \tag{5.3.22}$$

where

$$I_{00} := \int_S [N, [N, [E_0, N]]] ds = \int_S [N, E_0] ds = \int_D \nabla \times E_0 dx = O(a^3), \tag{5.3.23}$$

$$I_{01} := -\frac{\zeta}{i\omega\mu} \int_S [N, [N, \nabla \times E_0]] ds$$

$$= \frac{\zeta}{i\omega\mu} \int_S (\nabla \times E_0 - NN \cdot \nabla \times E_0) ds. \tag{5.3.24}$$

Let us note that this can be written as

$$I_{01} = \frac{\zeta|S|}{i\omega\mu} \tau \nabla \times E_0, \quad \tau := I - b, \tag{5.3.25}$$

where the matrices I and b can be written explicitly as follows:

$$I_{jm} = \delta_{jm} = \begin{cases} 0, & j \neq m, \\ 1, & j = m, \end{cases} \quad b_{jm} := \frac{1}{|S|} \int_S N_j(s) N_m(s) ds. \tag{5.3.26}$$

Note that if (5.3.17) holds, then

$$I_{01} = O(a^{2-\kappa}), \quad |I_{01}| \gg I_{00}. \tag{5.3.27}$$

Let us estimate I_2 and show that

$$I_2 \ll I_{01}. \tag{5.3.28}$$

One has

$$I_2 = I_{20} + I_{21}, \tag{5.3.29}$$

where

$$
\begin{aligned}
I_{21} &:= -\frac{\zeta}{i\omega\mu} \int_S \left[N, \left[N, -\nabla_x^2 \int_S g(x,t)J(t)dt \right] \right] ds \\
&= \frac{\zeta}{i\omega\mu} \int_S \left[N, \left[N, \int_S k^2 g(s,t)J(t)dt \right] \right] ds \\
&= O(|\zeta|a^3) \\
&\ll |I_{01}|. \tag{5.3.30}
\end{aligned}
$$

Here, the equation $-\nabla^2 g(x,t) = k^2 g(x,t) + \delta(x-t)$ was used, the term with $\delta(x-t)$ was set to zero since $x \neq t$, and then x was taken to $s \in S$.

Furthermore, the following relation:

$$\int_S |g(s,t)|dt = O(a), \quad a \to 0, \tag{5.3.31}$$

and the boundedness of J were used.

Let us estimate I_{20}:

$$I_{20} := -\frac{\zeta}{i\omega\mu} \int_S \left[N, \left[N, \nabla\nabla \cdot \int_S g(x,t)J(t)dt \right] \right] ds \tag{5.3.32}$$

One has

$$
\begin{aligned}
\nabla \cdot \int_S g(x,t)J(t)dt &= \int_S \nabla_x g(x,t) \cdot J(t)dt \\
&= -\int_S \nabla_t g(x,t) \cdot J(t)dt \\
&= \int_S g(x,t)\nabla_t \cdot J(t)dt, \tag{5.3.33}
\end{aligned}
$$

where an integration by parts on a closed surface S was done. Furthermore,

$$\nabla_x \int_S g(x,t) \nabla \cdot J(t) dt \bigg|_{x \to s} = - \int_S \nabla_t g(x,t) \nabla \cdot J(t) dt \bigg|_{x \to s}$$

$$= \int_S e_p g(x,t) \frac{\partial J_m(t)}{\partial t_p \partial t_m} dt \bigg|_{x \to s}, \quad (5.3.34)$$

where summation is understood over the repeated indices, $\{e_p\}_{p=1}^3$ is the Cartesian orthonormal basis of $\mathbb{R}^3$, twice continuous differentiability of $J(t)$ was assumed, and an integration by parts over the closed surface S was done.

From (5.3.32) and (5.3.34), one gets

$$|I_{20}| = O(|\zeta|a^3) \ll |I_{01}|. \quad (5.3.35)$$

Thus, from (5.3.30) and (5.3.35) it follows that

$$|I_2| \ll |I_{01}|. \quad (5.3.36)$$

An alternative proof of the estimate $I_2 = O(|\zeta a^3|)$ can be given as follows.

One has

$$\nabla_x \int_S g(x,t) \sigma(t) dt \bigg|_{x \to s^-} = \int_S \nabla_s g(s,t) \sigma(t) dt - \frac{\sigma(s)}{2} N_s,$$

$$\int_S ds \left[N, \left[N, \nabla \times \nabla \times \int_S g(x,t) J(t) dt \right] \right]$$

$$= \int_S ds \left[N, \left[N, \nabla \nabla \cdot \int_S g(x,t) J(t) dt - \nabla^2 \int_S g(x,t) J(t) dt \right] \right].$$

The last integral is

$$-\nabla^2 \int_S g(x,t) J(t) dt = k^2 \int_S g(x,t) J(t) dt$$

because $-\nabla^2 g(x,t) = k^2 g(x,t), x \neq t$.

The integral

$$\int_S |g(x,t)J(t)|dt = O(a), \quad a \to 0,$$

so

$$\int_S ds \left[N, \left[N, -\nabla^2 \int_S g(x,t)J(t)dt\right]\right] = O(a^3), \quad a \to 0.$$

The integral

$$\nabla\nabla \cdot \int_S g(x,t)J(t)dt = \nabla \int_S \nabla_x g(x,t) \cdot J(t)dt$$

$$= \nabla \int_S (-\nabla_t g(x,t) \cdot J(t))dt$$

$$= \nabla \int_S g(x,t)\nabla \cdot J(t)dt.$$

Consequently, denoting $\nabla \cdot J(t) = \sigma(t)$, one gets

$$\int_S ds \left[N, \left[N, \nabla \int_S g(x,t)\sigma(t)dt\right]\right]$$

$$= \int_S ds \left(N\frac{\partial}{\partial N^-} \int_S g(x,t)\sigma(t)dt - \nabla \int_S g(x,t)\sigma(t)dt\right)_{x \to s^-}$$

$$= \int_S ds \left(N_s \int_S \frac{\partial g(s,t)}{\partial N_s}\sigma(t)dt - N_s\frac{\sigma(s)}{2}\right.$$

$$\left. - \int_S \nabla_s g(s,t)\sigma(t)dt + N_s\frac{\sigma(s)}{2}\right)$$

$$= \int_S ds \int_S dt \left(N_s\frac{\partial g(s,t)}{\partial N_s} - \nabla_s g(s,t)\right)\sigma(t).$$

The expression under the parentheses is the derivative of $g(s,t)$ in a tangential to S direction. Therefore, the integral over t is bounded and the double integral is $O(a^3)$.

Let us estimate I_1. From formula (5.3.20) using formula (5.3.4), one obtains

$$I_1 = \int_S \int_S [N_s, [\nabla_s g(s,t), J(t)]] dt ds + \frac{1}{2} \int_S J(s) ds := I_{10} + I_{11}. \tag{5.3.37}$$

One has

$$I_{11} = \frac{Q}{2}, \tag{5.3.38}$$

and

$$I_{10} = \int_S ds \int_S \left(\nabla_s g(s,t) N_s \cdot J(t) - J(t) \frac{\partial g(s,t)}{\partial N_s} \right) dt. \tag{5.3.39}$$

It is known (see Chapter 11) that

$$-\int_S ds \frac{\partial g(s,t)}{\partial N_s} = \frac{1}{2} + o(1), \quad a \to 0. \tag{5.3.40}$$

Therefore,

$$-\int_S ds \int_S J(t) \frac{\partial g(s,t)}{\partial N_s} dt \sim \frac{Q}{2}, \quad a \to 0, \tag{5.3.41}$$

where the sign $\sim$ stands for asymptotically equal quantities.

Let us prove that

$$\left| \int_S ds \int_S \nabla_s g(s,t) N_s \cdot J(t) dt \right| \ll |I_{01}| = O(a^{2-\kappa}). \tag{5.3.42}$$

One has $N_t \cdot J(t) = 0$ because $J(t)$ is tangential to S. Thus,

$$|N_s \cdot J(t)| = |(N_s - N_t) \cdot J(t)| \le c|s - t|, \quad c = \text{const} > 0. \tag{5.3.43}$$

Consequently,

$$|\nabla_s g(s,t) N_s \cdot J(t)| = O\left(\frac{1}{|s - t|} \right), \quad \text{as } |s - t| \to 0. \tag{5.3.44}$$

Therefore,

$$\left| \int_S ds \int_S \nabla_s g(s,t) N_s \cdot J(t) dt \right| \leq c \int_S ds \int_S \frac{dt}{|s-t|} = O(a^3).$$

$$(5.3.45)$$

Estimates (5.3.45) implies (5.3.42).

Combining our estimates yields formula (5.3.15). Theorem 5.3.1 is proved. □

5.4 Summary of the results

The results of this chapter are formulas (5.3.8) and (5.3.15). These formulas solve the problem of electromagnetic wave scattering by a small impedance body of an arbitrary shape.

Chapter 6

Many-Body Scattering Problem in the Case of Small Scatterers

6.1 Reduction of the problem to linear algebraic system

Consider electromagnetic (EM) wave scattering problem by many small impedance bodies.

This problem is as follows:

$$\nabla \times E = i\omega\mu H, \quad \text{in } D' := \mathbb{R}^3 \setminus D, \quad D := \bigcup_{m=1}^{M} D_m, \qquad (6.1.1)$$

$$\nabla \times H = -i\omega\epsilon' E, \quad \text{in } D', \qquad (6.1.2)$$

$$[N, [E, N]] = \zeta_m[N, H], \quad \text{on } S_m, \quad 1 \leq m \leq M, \qquad (6.1.3)$$

$$E = E_0 + v, \qquad (6.1.4)$$

where v satisfies the radiation condition (5.1.6) and E_0 is the incident plane wave (5.1.5).

Let us assume that

$$\zeta_m = \frac{h(x_m)}{a^\kappa}, \quad 0 \leq \kappa < 1, \quad x_m \in D_m, \quad 1 \leq m \leq M, \qquad (6.1.5)$$

where $h(x)$ is a continuous function in a bounded domain Ω, $D \subset \Omega$,

$$\text{Re}h(x) \geq 0, \quad \text{Im}\epsilon' = \frac{\sigma}{\omega} \geq 0, \tag{6.1.6}$$

$$\epsilon' = \epsilon_0, \quad \mu = \mu_0 \quad \text{in } \Omega' := \mathbb{R}^3 \setminus \Omega. \tag{6.1.7}$$

Let us assume that

$$\mathcal{N}(\Delta) = \frac{1}{a^{2-\kappa}} \int_\Delta N(x)dx[1 + o(1)], \quad a \to 0, \tag{6.1.8}$$

where $\mathcal{N}(\Delta)$ is the number of small bodies in Δ, Δ is an arbitrary open subset of Ω,

$$N(x) \geq 0, \quad N(x) \in C(\Omega), \tag{6.1.9}$$

and $\kappa \in [0,1)$ is the parameter from (6.1.5).

The smoothness assumption on the surface $S = \bigcup_{m=1}^M S_m$ are the same as before, $S \in C^{1,\gamma}, \gamma \in (0,1]$.

If E is found, then

$$H = \frac{\nabla \times E}{i\omega\mu}, \tag{6.1.10}$$

and E solves the equation

$$\nabla \times \nabla \times E = k^2 E, \quad k^2 = \omega^2 \epsilon' \mu, \tag{6.1.11}$$

which holds if $\mu =$const, which we assume. If (6.1.10) and (6.1.11) hold, then equation (6.1.2) holds. Thus, our scattering problem is reduced to finding E solving (6.1.11), satisfying condition (6.1.3) in the form

$$[N,[E,N]] - \frac{\zeta}{i\omega\mu}[N, \nabla \times E] = 0 \quad \text{on } S_m, \quad 1 \leq m \leq M, \tag{6.1.12}$$

and condition (6.1.4).

As in Chapter 5, one proves the existence and uniqueness of the solution to problems (6.1.1)–(6.1.4).

Theorem 6.1.1. *Problems (6.1.1)–(6.1.4) have a unique solution under the assumption (6.1.6).*

We look for the solution of the form

$$E(x) = E_0(x) + \sum_{m=1}^{M} \nabla \times \int_{S_m} g(x, t) J_m(t) dt. \qquad (6.1.13)$$

This equation can be rewritten as

$$E(x) = E_0(x) + \sum_{m=1}^{M} [\nabla g(x, x_m), Q_m] + \sum_{m=1}^{M} \nabla$$

$$\times \int_{S_m} (g(x, t) - g(x, x_m)) J_m(t) dt$$

$$:= E_0(x) + I_1 + I_2, \quad x_m \in D_m, \qquad (6.1.14)$$

where

$$Q_m := \int_{S_m} J_m(t) dt. \qquad (6.1.15)$$

If the bodies D_m are small, then

$$|I_2| \ll |I_1|, \quad a \to 0. \qquad (6.1.16)$$

Let us explain formula (6.1.16). The term

$$[\nabla g(x, x_m), Q_m] = O\left(\max\left(\frac{1}{d^2}, \frac{k}{d}\right)\right) |Q_m|, \qquad (6.1.17)$$

where $d = |x - x_m|$, and

$$|\nabla g(x, x_m)| = O\left(\max\left(\frac{1}{d^2}, \frac{k}{d}\right)\right). \qquad (6.1.18)$$

The term

$$\left| \nabla \times \int_{S_m} (g(x, t) - g(x, x_m)) J_m(t) dt \right| = O\left(\max\left(\frac{a}{d^3}, \frac{k^2 a}{d}\right)\right) |Q_m|, \qquad (6.1.19)$$

because

$$|\nabla_x (g(x, t) - g(x, x_m))| = O\left(\max\left(\frac{a}{d^3}, \frac{k^2 a}{d}\right)\right). \qquad (6.1.20)$$

The ratio

$$\left|\frac{I_2}{I_1}\right| = O\left(\frac{a}{d} + ka\right) \to 0, \quad a \to 0. \tag{6.1.21}$$

This explains estimate (6.1.16), which justifies the asymptotic equation

$$E(x) \sim E_0(x) + \sum_{m=1}^{M} [\nabla g(x, x_m), Q_m], \quad a \to 0, \tag{6.1.22}$$

where *the ~ sign stands for quantities asymptotically equal as $a \to 0$.* Formula (6.1.22) is of fundamental importance because it reduces the solution to the scattering problems (6.1.1)–(6.1.4) to finding quantities $Q_m, 1 \leq m \leq M$, rather than to finding the functions $J_m(t)$, $1 \leq m \leq M$. This is a drastic simplification of the scattering problem, and it allows one to solve the problem numerically when M is very large.

The remaining part of the solution consists of finding $Q_m, 1 \leq m \leq M$. To find Q_m, let us introduce the effective field acting on the j-th body

$$E_e(x) = E_0(x) + \sum_{m \neq j} [\nabla g(x, x_m), Q_m]. \tag{6.1.23}$$

Using formula (5.3.15) and replacing E_0 in this formula by the effective field, one gets

$$Q_m = -\frac{\zeta_m |S_m|}{i\omega\mu} \tau_m \nabla \times E_{em}, \quad 1 \leq m \leq M. \tag{6.1.24}$$

Let us assume that

$$\zeta_m = \frac{h(x_m)}{a^\kappa} := \frac{h_m}{a^\kappa} \quad 0 \leq \kappa < 1, \quad \text{Re}h \geq 0, \tag{6.1.25}$$

$h(x) \in C(\Omega)$ is a continuous function,

$$|S_m| = ca^2, \quad c = \text{const} > 0, \tag{6.1.26}$$

$$\tau_m = \tau = I - b, \tag{6.1.27}$$

where the matrix b is defined in (5.3.16), $|S_m|$ is the surface area of S_m, and our assumptions (6.1.26) and (6.1.27) mean that all the small bodies have the same size, shape and orientation. The theory can be easily generalized to the case when $c = c_m$ and $\tau = \tau_m$, that is, to the case of small bodies of various shapes. From formulas (6.1.23)–(6.1.27), it follows that

$$E_e(x) = E_0(x) - \frac{c}{i\omega\mu} \sum_{m \neq j} [\nabla g(x, x_m), \tau\nabla \times E_{em}] h_m a^{2-\kappa}, \quad (6.1.28)$$

and τ is defined in (5.3.16), where $S = S_m$.

In formula (6.1.28), the vectors E_{em} are unknown. Problems (6.1.1)–(6.1.4) are solved if one finds these vectors. Let us derive a linear algebraic system for finding these vectors. Denote $\nabla \times E_{em} := A_m$. Apply the operation $\nabla\times$ to the equation (6.1.28) and set $x = x_j$ afterwards. This yields

$$A_j = A_{0j} - \frac{ca^{2-\kappa}}{i\omega\mu} \sum_{m \neq j} k^2 g(x_j, x_m) \tau A_m h_m$$

$$+ (\tau A_m \cdot \nabla_x)\nabla g(x, x_m)|_{x=x_j} h_m, \quad (6.1.29)$$

where $1 \leq j \leq M$. Here we have used the known formula for $A = A(x)$ and B

$$\nabla \times [A, B] = (B \cdot \nabla)A - B\nabla \cdot A, \quad (6.1.30)$$

where B is a constant vector,

$$A = \nabla g(x, x_m), \quad B = \tau A_m, \quad (6.1.31)$$

and took into account that B does not depend on x so that

$$-\tau A_m \nabla \cdot \nabla g(x, x_m) = -\tau A_m \nabla^2 g(x, x_m)|_{x=x_j}.$$

Since

$$-\nabla^2 g(x, x_m)|_{x=x_j} = k^2 g(x_j, x_m), \quad x_j \neq x_m, \quad (6.1.32)$$

one gets (6.1.29). From the linear algebraic system (LAS) (6.1.29), one finds $A_m = \nabla \times E_{em}$ and then one calculates the effective field $E_e(x)$ by formula (6.1.28).

This completes the description of our numerical method for solving the scattering problems (6.1.1)–(6.1.4).

The effective field is asymptotically equal to the field $E(x)$, defined by formula (6.1.22), because the input of the field $[\nabla g(x, x_j), Q_j]$ is

$$O\left(a^{2-\kappa} \max\left(\frac{1}{d^2}, \frac{k}{d}\right)\right) \to 0, \quad a \to 0, \tag{6.1.33}$$

if $\frac{a^{1-\kappa/2}}{d} \to 0$ as $a \to 0$.

Let us summarize the result.

Theorem 6.1.2. *The solution to the scattering problems (6.1.1)–(6.1.4) is given by formula (6.1.28) in which vectors $A_m := \nabla \times E_{em}$ are found from linear algebraic system (6.1.29).*

6.2   Derivation of the integral equation for the effective field

In this section, we keep the assumptions made in Section 6.1, in particular the assumption (6.1.8). From this assumption it follows that $d = O(a^{\frac{2-\kappa}{3}})$, where d is the minimal distance between neighboring particles (bodies).

Indeed, the total number of small bodies is $O(\frac{1}{d^3})$ and it is equal to $O(\frac{1}{a^{2-\kappa}})$ by formula (6.1.8). This yields the estimate $d = O(a^{\frac{2-\kappa}{3}})$.

To derive the integral equation for the limiting effective field as $a \to 0$, one starts with formula (6.1.28). Let us partition the bounded domain Ω into a union of small cubes $\Delta_p, 1 \le p \le P$, with the side $b = b(a) \to 0$ as $a \to 0$,

$$a \ll d \ll b, \quad \Omega = \bigcup_{p=1}^{P} \Delta_p. \tag{6.2.1}$$

The cubes Δ_p and Δ_q may have common parts of the boundaries but have no common interior points. In each cube Δ_p, we choose a point x_p. The number P of these points is much smaller than the number M of all points $x_m, 1 \le m \le M$. Recall that the functions $h(x), N(x)$ are assumed continuous, and the function $E_e(x)$ is assumed continuously differentiable.

The sum in (6.1.28) can be written as

$$\sum_{m \neq j} [\nabla g(r, r_m), \tau \nabla \times E_{em}] h_m a^{2-\kappa}$$

$$= \sum_{p \neq q} [\nabla g(x, x_p), \tau \nabla \times E_{ep}] h_p a^{2-\kappa} \sum_{x_m \in \Delta_p} 1$$

$$= \sum_{p \neq q} [\nabla g(x, x_p), \tau \nabla \times E_{ep}] h_p a^{2-\kappa} |\Delta_p|. \qquad (6.2.2)$$

Here the formula

$$a^{2-\kappa} \sum_{x_m \in \Delta_p} 1 = \int_{\Delta_p} N(x) dx (1 + o(1)) = N(x_p) |\Delta_p|, \quad a \to 0,$$

$$(6.2.3)$$

was used, which is justified because the diameter of Δ_p tends to zero as $a \to 0$.

By this reason we have set

$$\nabla g(x, x_m) = \nabla g(x, x_p), \quad \nabla \times E_{em} = \nabla \times E_{ep}, \quad h_m = h_p, \quad x \notin \Delta_p,$$

for all points in $x_m \in \Delta_p$. By the continuity of these functions, the errors of these equations are negligible as $a \to 0$. The sum on the right-hand side of (6.2.2) is the Riemannian sum for the integral

$$\int_\Omega [\nabla_x g(x, y), \tau \nabla \times E(y)] h(y) N(y) dy, \qquad (6.2.4)$$

and the limiting form of the equation (6.1.28) is the integral equation

$$E(x) = E_0(x) - \frac{c}{i\omega\mu} \int_\Omega [\nabla_x g(x, y), \tau \nabla \times E(y)] h(y) N(y) dy. \quad (6.2.5)$$

This is the integral equation for the limiting field E. The constant c in (6.2.5) is defined by the relation $|S| = ca^2$. If all the small bodies are balls of radius a, then $c = 4\pi$.

Let us formulate the result.

Theorem 6.2.1. *The limiting effective field satisfies equation* (6.2.5).

The integral equation (6.2.5) can be written as

$$E(x) = E_0(x) - \frac{c}{i\omega\mu}\nabla \times \int_\Omega g(x,y)\tau\nabla \times E(y)h(y)N(y)dy. \quad (6.2.6)$$

6.3   Summary of the results

The main results of this chapter are formulas (6.1.28), (6.1.29), and (6.2.6).

Chapter 7

Creating Materials with a Desired Refraction Coefficient

7.1 A formula for the refraction coefficient

In the original material the refraction coefficient is $n_0 = \sqrt{\frac{\epsilon' \mu}{\epsilon_0 \mu_0}}$. In the material one obtained in the limit $a \to 0$ after embedding many small particles in the original material, the refraction coefficient is different from the original one. To find it, let us apply the operator $\nabla \times \nabla \times = \nabla \nabla \cdot - \nabla^2$ to equation (6.2.6). Since $\nabla \cdot \nabla \times A \equiv 0$, and

$$-\nabla^2 g(x, y) = k^2 g(x, y) + \delta(x - y), \qquad (7.1.1)$$

one gets

$$\nabla \times \nabla \times E(x) = \nabla \times \nabla \times E_0(x) - \frac{c}{i\omega\mu}(\nabla\nabla \cdot - \nabla^2)\nabla$$

$$\times \int_\Omega g(x, y)\tau\nabla \times E(y)h(y)N(y)dy$$

$$= k^2 E_0(x) - \frac{c}{i\omega\mu}\nabla \times \int_\Omega \left(k^2 g(x, y) + \delta(x - y)\right)\tau\nabla$$

$$\times E(y)h(y)N(y)dy$$

$$= k^2 E(x) - \frac{c}{i\omega\mu}\nabla \times (\tau\nabla \times E(x)h(x)N(x)).$$

$$(7.1.2)$$

Let us assume that τ is proportional to the identity matrix

$$\tau = c_\tau I, \quad c_\tau = \text{const.}$$

This, for example, is the case when the small bodies are balls of radius a. The constant $c_\tau = 2/3$ in this case because $\tau_{jm} = \delta_{jm} - b_{jm}$, where for balls one has

$$b_{jm} = \frac{1}{4\pi a^2} \int_{|s|=a} N_j(s)N_m(s)ds = \frac{1}{3}\delta_{jm}, \quad \delta_{jm} = \begin{cases} 1, & j = m, \\ 0, & j \neq m. \end{cases}$$
$$(7.1.3)$$

If $\tau = c_\tau I$, then formula (7.1.2) takes the form

$$\nabla \times \nabla \times E(x) = k^2 E(x) - \frac{cc_\tau}{i\omega\mu} h(x)N(x)\nabla \times \nabla \times E$$

$$- \frac{cc_\tau}{i\omega\mu}[\nabla(h(x)N(x)), \nabla \times E]. \quad (7.1.4)$$

Therefore,

$$\nabla \times \nabla \times E(x) = \frac{k^2 E(x)}{1 + \frac{cc_\tau}{i\omega\mu}h(x)N(x)} - \frac{cc_\tau}{i\omega\mu}\frac{[\nabla(h(x)N(x)), \nabla \times E]}{1 + \frac{cc_\tau}{i\omega\mu}h(x)N(x)}.$$
$$(7.1.5)$$

To interpret this equation physically, let us consider Maxwell's equations

$$\nabla \times E = i\omega\mu H, \quad \nabla \times H = -i\omega\epsilon' E \quad (7.1.6)$$

and apply the curl to the first equation assuming that $\mu = \mu(x)$. This yields

$$\nabla \times \nabla \times E = k^2 E + \left[\frac{\nabla\mu(x)}{\mu(x)}, \nabla \times E\right] \quad (7.1.7)$$

Comparing (7.1.7) with (7.1.5) we can conclude that the new medium, described by equation (7.1.5), has *new refraction coefficient*

$$n(x) = \frac{n_0}{\sqrt{1 + \frac{cc_\tau}{i\omega\mu}h(x)N(x)}}, \quad n_0 = \sqrt{\frac{\epsilon'\mu}{\epsilon_0\mu_0}} \quad (7.1.8)$$

and *new magnetic permeability*

$$\mu(x) := \left(1 + \frac{cc_T}{i\omega\mu}h(x)N(x)\right)^{-1}, \qquad (7.1.9)$$

where

$$\frac{\nabla\mu(x)}{\mu(x)} = -\frac{cc_T}{i\omega\mu}\frac{\nabla(h(x)N(x))}{1 + \frac{cc_T}{i\omega\mu}h(x)N(x)}. \qquad (7.1.10)$$

7.2 Formula for magnetic permeability

This formula is formula (7.1.9) derived in the previous section.

Let us discuss the recipe for creating a desired refraction coefficient and a desired magnetic permeability.

Let us rewrite formulas (7.1.8) and (7.1.9) as follows:

$$n(x) = \frac{n_0}{\left(1 - i\frac{cc_T}{\omega\mu}N(x)h(x)\right)^{1/2}}, \qquad (7.2.1)$$

$$\mu(x) = \frac{1}{1 - i\frac{cc_T}{\omega\mu}N(x)h(x)}. \qquad (7.2.2)$$

The constant $\frac{cc_T}{\omega\mu} > 0$ is known. The functions $N(x) \geq 0$ and $h(x), h_1 = \mathrm{Re}h(x) \geq 0, h = h_1 + ih_2$, are at the disposal of the experimenter.

Formulas (7.2.1) and (7.2.2) can be written as

$$n(x) = n_0(x)\left(1 + \frac{cc_T N(x)h_2(x)}{\omega\mu} - i\frac{cc_T N(x)h_1(x)}{\omega\mu}\right)^{-\frac{1}{2}}, \qquad (7.2.3)$$

$$\mu(x) = \left(1 + \frac{cc_T N(x)h_2(x)}{\omega\mu} - i\frac{cc_T N(x)h_1(x)}{\omega\mu}\right)^{-1}. \qquad (7.2.4)$$

If one is interested only in the refraction coefficient, then by choosing $h = h_1 + ih_2, h_1 \geq 0$, and $N(x) \geq 0$, one can create any desired refraction coefficient by formula (7.2.3). Indeed, suppose that $n_0 > 0$

and a desired $n(x) = n_1(x) + in_2(x)$. Denote $\frac{cc_\tau}{\omega\mu} := \psi > 0$. Then

$$n_1(x) + in_2(x) = (1 + \psi N(x)h_2(x) - i\psi N(x)h_1(x))^{-\frac{1}{2}} \qquad (7.2.5)$$

There are no physical restrictions on the $h_2(x) \in (-\infty, \infty)$. In applications to creating metamaterials, much interest is in creating materials with negative refraction coefficient. If $h_1(x) > 0$ and $h_2(x) > 0$, then for small $h_1(x)$ the argument of the expression under the square root in formula (7.2.4) is close to 2π and less than 2π. Therefore, the square root has the argument close to π and less than π. This means that the real part of $n(x)$ is negative.

7.3 Summary of the results

The main results of this chapter are formulas (7.2.3)–(7.2.4). These formulas give a recipe for creating materials with a desired refraction coefficient by choosing $N(x)$ and $h(x) = h_1(x) + ih_2(x)$, where the functions $N(x) \geq 0, h_1(x) \geq 0$, and $h_2(x)$ are at the disposal of the experimenter.

The technological problem of practical importance is to create a small impedance particle with a desired boundary impedance.

Chapter 8

Electromagnetic Wave Scattering by Many Nanowires

8.1 Statement of the problem

In this chapter, electromagnetic wave scattering by many thin, parallel to z-axis, infinite cylinders, on the surfaces of which impedance boundary conditions are imposed, is studied. Let D_m denote the crosssection of the m-th cylinder, a be its characteristic size $a = \frac{1}{2}\text{diam}D_m$. We do not assume that D_m is a disc. The shape of D_m is arbitrary. For simplicity, we assume that all the cylinders are identical, but this assumption is not necessary for the validity of our theory. Let $t_m \in D_m$ be an arbitrary point inside D_m. If D_m are discs, we take as x_m their centers. By $M = M(a)$, we denote the total number of thin cylinders. The points x_m are distributed in a bounded domain Ω on a plane perpendicular to the z-axis. This plane is the x–y plane. These cylinders model nanowires. They are thin in the sense that $ka \ll 1$, where k is the wave number. The distribution of the points t_m, and therefore the cylinders, is given by the following law:

$$\mathcal{N}(\Delta) = \frac{1}{a^{1-\kappa}} \int_{\Delta} N(\hat{x}) d\hat{x} \, [1 + o(1)], \quad a \to 0, \qquad (8.1.1)$$

where $0 \le \kappa < 1$ is a given number, $c_s a = |S|$ is the length of the boundary of the crosssection of a cylinder, and we assumed for simplicity that all the cylinders are identical, $c =$ const,

$$\mathcal{N}(\Delta) = \sum_{t_m \in \Delta} 1 \tag{8.1.2}$$

is the number of the points t_m in an arbitrary open subset of the plane domain Ω, $\hat{x} = (x_1, x_2) \in \Omega$ is a point, $N(x) \geq 0$ is a given continuous function. We assume that the domains D_m are non-intersecting and denote by S_m their boundaries. By $D = \bigcup_{m=1}^{M} D_m$, we denote the union of D_m, $S = \bigcup_{m=1}^{M} S_m$, by C_m, the cylinder $D_m \times (-\infty, \infty)$, by $C = \bigcup_{m=1}^{M} C_m$, and by $C' := \mathbb{R} \backslash C$. Let P denote the x–y plane, $D' := P \backslash D$, the complement of D in P.

The electromagnetic (EM) wave scattering problem is formulated as follows:

$$\nabla \times E = i\omega\mu H \tag{8.1.3}$$

$$\nabla \times H = -i\omega\epsilon E, \tag{8.1.4}$$

these Maxwell equations hold in C', ω is the frequency, ϵ and μ are constant parameters, permittivity and permeability. On the boundary S of the cylinders, the impedance boundary condition holds

$$E_t = \zeta_m[N, H] \quad \text{on } C_m, \quad 1 \leq m \leq M, \tag{8.1.5}$$

where E_t is the tangential component of E, ζ_m is the boundary impedance, $\text{Re}\zeta_m \geq 0$, N is the unit normal to C_m pointing out of C_m, H is the magnetic field. We define

$$E_t := [N, [E, N]] = E - NN \cdot E. \tag{8.1.6}$$

By $[E, H] := E \times H$, the vector product is denoted, and $E \cdot H$ stands for the scalar product of two vectors. Note that

$$[N, H] = [N, H_t]. \tag{8.1.7}$$

The quantity

$$k = \omega\sqrt{\epsilon\mu}, \quad n_0 = \sqrt{\epsilon\mu} \tag{8.1.8}$$

is the wave number and n_0 is the refraction coefficient. We are looking for the solution to equations (8.1.3)–(8.1.5) of the form

$$E(x) = E_0(x) + V(x), \quad x := (\hat{x}, z), \tag{8.1.9}$$

where E_0 is the incident field, $V = e^{ik_3z}v$ is the scattered field satisfying the radiation condition

$$\sqrt{r}\left(\frac{\partial v}{\partial r} - ikv\right) = o(1), \quad r \to \infty, \quad r := |\hat{x}|. \tag{8.1.10}$$

We assume that

$$E_0(x) = \frac{e^{i\kappa y + ik_3 z}}{k}(-k_3 e_2 + \kappa e_3), \quad \kappa^2 + k_3^2 = k^2. \tag{8.1.11}$$

Here $\{e_m\}_{m=1}^3$ is a Cartesian basis of $\mathbb{R}^3$, e_3 is along z-axis, e_1, e_2 are along x and y axes.

Let us consider solutions with $H_z = H_3 = 0$, that is, E-waves (or TH-waves). For these waves

$$E = \sum_{j=1}^3 E_j e_j, \quad H = H_1 e_1 + H_2 e_2 = \frac{\nabla \times E}{i\omega\mu}. \tag{8.1.12}$$

Note that it is sufficient to find E since H is calculated by the formula $H = \frac{\nabla \times E}{i\omega\mu}$. Let us derive formulas for E as follows:

$$E_j = \frac{ik_3}{\kappa^2} u_{x_j} e^{ik_3 z}, \quad j = 1, 2; \quad E_3 = u e^{ik_3 z}, \tag{8.1.13}$$

where $u = u(x, y)$ solves the following problem:

$$(\nabla^2 + \kappa^2)u = 0 \quad \text{in } D', \quad \kappa^2 = k^2 - k_3^2, \tag{8.1.14}$$

$$u_N + i\xi u = 0 \quad \text{on } \partial D, \quad \xi := \frac{\omega\mu\kappa^2}{\zeta k^2}, \tag{8.1.15}$$

$$u = e^{i\kappa y} + w, \tag{8.1.16}$$

and w satisfies the radiation condition (8.1.10). To derive formula (8.1.13), let us look for the solution to equations (8.1.3)–(8.1.4) of the form

$$E_1(x) = e^{ik_3 z} \tilde{E}_1(\hat{x}), \quad E_2(x) = e^{ik_3 z} \tilde{E}_2(\hat{x}), \quad E_3(x) = e^{ik_3 z} u(\hat{x}),$$
$$(8.1.17)$$

$$H_1(x) = e^{ik_3 z} \tilde{H}_1(\hat{x}), \quad H_2(x) = e^{ik_3 z} \tilde{H}_2(\hat{x}), \quad H_3 = 0. \quad (8.1.18)$$

Let $\partial_j := \frac{\partial}{\partial x_j}$. Equation (8.1.3) yields

$$\partial_2 u - ik_3 \tilde{E}_2 = i\omega\mu\tilde{H}_1, \tag{8.1.19}$$

$$-\partial_1 u + ik_3 \tilde{E}_1 = i\omega\mu\tilde{H}_2, \tag{8.1.20}$$

$$\partial_1 \tilde{E}_2 - \partial_2 \tilde{E}_1 = 0. \tag{8.1.21}$$

Equation (8.1.4) yields

$$ik_3 \tilde{H}_2 = i\omega\epsilon\tilde{E}_1, \tag{8.1.22}$$

$$ik_3 \tilde{H}_1 = -i\omega\epsilon\tilde{E}_2, \tag{8.1.23}$$

$$\partial_1 \tilde{H}_2 - \partial_2 \tilde{H}_1 = -i\omega\epsilon u. \tag{8.1.24}$$

Let us eliminate $\tilde{H}_1$ and $\tilde{H}_2$ from equations (8.1.19)–(8.1.21) using equations (8.1.22)–(8.1.24). This yields formula (8.1.13) as follows:

$$\tilde{E}_1 = \frac{ik_3}{\kappa^2} u_x, \quad \tilde{E}_2 = \frac{ik_3}{\kappa^2} u_y, \quad \tilde{E}_3 = u. \tag{8.1.25}$$

Applying the operator ∂_2 to (8.1.19), ∂_1 to (8.1.20), and subtracting from the first equation the second, one gets

$$\partial_1^2 u + \partial_2^2 u - ik_3(\partial_1 \tilde{E}_1 + \partial_2 \tilde{E}_2) = i\omega\mu(\partial_2 \tilde{H}_1 - \partial_1 \tilde{H}_2). \tag{8.1.26}$$

Using equation (8.1.24) yields

$$i\omega\mu(\partial_2 \tilde{H}_1 - \partial_1 \tilde{H}_2) = -\omega^2 \epsilon\mu u = -k^2 u. \tag{8.1.27}$$

Applying ∂_1 to equation (8.1.22), ∂_2 to equation (8.1.23), subtracting from the first equation the second, and using (8.1.24) yields the following:

$$\partial_1 \tilde{E}_1 + \partial_2 \tilde{E}_2 = \frac{k_3}{\omega\epsilon}(\partial_1 \tilde{H}_2 - \partial_2 \tilde{H}_1) = -ik_3 u. \tag{8.1.28}$$

From (8.1.26)–(8.1.28), one gets equation (8.1.14) with $\kappa^2 = k^2 - k_3^2$.

The impedance boundary condition (8.1.5) yields

$$E - NE \cdot N = \zeta e_3(N_1 H_2 - N_2 H_1), \qquad (8.1.29)$$

because $N_3 = 0$ and $H_3 = 0$. From (8.1.29), one obtains

$$E_3 = \zeta(N_1 H_2 - N_2 H_1)$$

$$= \frac{\zeta}{i\omega\mu}(N_1(\partial_3 E_1 - \partial_1 E_3) - N_2(\partial_2 E_3 - \partial_3 E_2)), \qquad (8.1.30)$$

or

$$\frac{i\omega\mu}{\zeta}E_3 = -N_1\partial_1 E_3 - N_2\partial_2 E_3 + N_1\partial_3 E_1 + N_2\partial_3 E_2. \qquad (8.1.31)$$

This and (8.1.25) yield $i\xi u + u_N(1 + \frac{k_3^2}{\kappa^2}) = 0$ on ∂D, where $\xi := \frac{\omega\mu}{\zeta}$. If $\kappa = 0$ then this boundary condition reduces to $u_N = 0$ on ∂D. The case $k_3 = 0$ is considered below.

To have boundary condition (8.1.31) imposed only on E_3, one requires $E_1 = E_2 = 0$. Then (8.1.31) takes the form

$$\frac{\partial E_3}{\partial N} + \frac{i\omega\mu}{\zeta}E_3 = 0, \qquad (8.1.32)$$

or, taking into account (8.1.17),

$$\frac{\partial u}{\partial N} + \frac{i\omega\mu}{\zeta}u = 0. \qquad (8.1.33)$$

If $E_1 = E_2 = 0$, then $k_3 = 0$, $\kappa^2 = k^2$, and formula (8.1.13) become

$$E_1 = E_2 = 0, \quad E_3 = u(\hat{x}), \qquad (8.1.34)$$

$$H = \frac{\nabla \times E}{i\omega\mu}, \qquad (8.1.35)$$

where u solves the problem

$$(\partial_1^2 + \partial_2^2 + k^2)u = 0 \quad \text{in } D', \qquad (8.1.36)$$

$$u_N + i\xi u = 0, \quad \xi := \frac{\omega\mu}{\zeta}, \qquad (8.1.37)$$

$\mathrm{Re}\,\xi \geq 0$ since $\mathrm{Re}\,\zeta \geq 0$ and $\omega\mu > 0$, and (8.1.11) yields

$$u = e^{iky} + w, \qquad (8.1.38)$$

where w satisfies the radiation condition. If u is found as the solution to (8.1.36)–(8.1.38), then E is found by formula (8.1.34) and H is found by formula (8.1.35).

8.2 Asymptotic solution of the problem

The aim of this section is to find asymptotic behavior of u as $a \to 0$, $M = M(a) \to \infty$, and to derive the equation for the limiting field in the new limiting medium.

Let us start with problems (8.1.36)–(8.1.38), for one small body $D = D_1$. Let $x_1 \in D_1$ be an arbitrary point inside D_1, $a = \frac{1}{2}\text{diam}D_1$, and the smallness of D_1 means that

$$ka \ll 1. \tag{8.2.1}$$

Let us look for the solution to problems (8.1.36)–(8.1.38) of the form

$$u(\hat{x}) = e^{iky} + \int_{S_1} g(\hat{x}, t)\sigma(t)dt, \tag{8.2.2}$$

$$g(\hat{x}, t) = \frac{i}{4}H_0^1(k|\hat{x} - t|), \tag{8.2.3}$$

where $H_0^1(z)$ is the Hankel function of order 1 with index 0, and σ is unknown. For any σ, function (8.2.2) solves equation (8.1.36) and satisfies condition (8.1.38). Problems (8.1.36)–(8.1.38) have a unique solution. If σ can be found from the boundary condition (8.1.37), then formula (8.2.2) gives the solution to problems (8.1.36)–(8.1.38).

Before using the boundary condition (8.1.37) let us recall some known facts about the function $g(kr) = \frac{i}{4}H_0^1(kr)$, see, for example, Lebedev (1972):

$$g(kr) = \alpha(k) + \frac{1}{2\pi}\ln\frac{1}{r} + o(1), \quad r \to 0, \tag{8.2.4}$$

$$\alpha := \alpha(k) := \frac{i}{4} + \frac{1}{2\pi}\ln\frac{2}{k}, \tag{8.2.5}$$

$$g(kr) = \frac{i}{4}\sqrt{\frac{2}{\pi kr}}e^{i\left(kr - \frac{\pi}{4}\right)}(1 + o(1)), \quad r \to \infty. \tag{8.2.6}$$

From (8.2.2) for $|\hat{x} - t| \gg a$, one obtains

$$u(\hat{x}) = e^{iky} + g(\hat{x}, t_1)Q \qquad (8.2.7)$$

$$Q := \int_{S_1} \sigma(t)dt, \qquad (8.2.8)$$

where $t_1 \in D_1$ is an arbitrary point.

In (8.2.7), we neglected the term

$$J_2 := \int_{S_1} [g(\hat{x}, t) - g(\hat{x}, t_1)]\sigma(t)dt \qquad (8.2.9)$$

compared with

$$J_1 = g(\hat{x}, t_1)Q.$$

As $a \to 0$, one has

$$|J_2| \ll |J_1|.$$

The basic advantage of our method for solving problems (8.1.36)–(8.1.38) under the smallness assumption (8.2.1) is the reduction of this problem to finding *just one number* Q, defined in (8.2.8), rather than to finding *an unknown function* $\sigma(t)$. For the number Q, we will derive an explicit analytical formula asymptotically exact as $a \to 0$. To do this, let us use the exact boundary condition (8.1.37). Let

$$g_0(s, t) := \frac{1}{2\pi} \ln \frac{1}{r_{st}}, \quad r_{st} := |s - t|; \qquad (8.2.10)$$

$$u_0 := e^{iks_2}; \quad A\sigma := \int_{S_1} \frac{\partial g_0(s, t)}{\partial N_s} \sigma(t)dt. \qquad (8.2.11)$$

We will use the following formula (see Chapter 11 for a proof of this formula):

$$\frac{\partial}{\partial N_s^-} \int_{S_1} g_0(s, t)\sigma(t)dt = \frac{A\sigma - \sigma}{2}, \qquad (8.2.12)$$

where the limiting value of the normal derivative of the single layer potential is taken from outside D.

Condition (8.1.37) yields

$$u_{0N} + i\xi u_0 + \frac{A\sigma - \sigma}{2} + i\xi\alpha Q + i\xi \int_{S_1} g_0(s,t)\sigma(t)dt = 0. \quad (8.2.13)$$

Here we have used for small $|s - t|$ the following formula:

$$u = u_0 + \alpha Q + \int_{S_1} g_0(s,t)\sigma(t)dt, \quad (8.2.14)$$

where α is defined in formula (8.2.5). Formula (8.2.14) is valid up to terms of higher order of smallness.

If $ka \ll 1$, then $u_0 \sim 1$, where the sign $\sim$ means equality up to the terms of higher order of smallness, and $u_{0N} \sim ikN_2$.

Let us integrate formula (8.2.13) over S_1 and keep the main terms as $a \to 0$. The first term is

$$\int_{S_1} u_{0N} ds = \int_{D_1} \Delta u_0 d\hat{x} = O(a^2). \quad (8.2.15)$$

The next term is

$$i\xi \int_{S_1} u_0 ds \sim i\xi |S_1| u_0(x_1), \quad (8.2.16)$$

where $|S_1|$ is the length of the boundary S_1.

In Chapter 11, one can find the formula

$$\int_S A\sigma ds = -\int_S \sigma ds.$$

This implies the following relation:

$$\int_{S_1} \frac{A\sigma - \sigma}{2} ds = -Q. \quad (8.2.17)$$

The term

$$\int_{S_1} i\xi\alpha Q ds = O(a)|\xi||Q| \ll |Q|, \quad (8.2.18)$$

if $a\xi \to 0$ as $a \to 0$. Thus, one may neglect this term as $a \to 0$.

The last term is

$$i\xi \int_{S_1} ds \int_{S_1} g_0(s,t)\sigma(t)dt = \int_{S_1} dt\sigma(t)i\xi \int_{S_1} ds g_0(s,t) = o(Q).$$

$$(8.2.19)$$

So, this term is also negligible as $a \to 0$.

Here we have used the estimate

$$\int_{S_1} g_0(s,t)ds = O(a|\ln a|), \quad a \to 0. \tag{8.2.20}$$

Collecting the main terms yields the following formula for Q:

$$Q = i\xi|S_1|u_0(t_1), \quad \xi = \frac{\omega\mu}{\zeta}, \tag{8.2.21}$$

where $|S_1|$ is the length of the contour S_1. Consequently, the solution to the scattering problems (8.1.36)–(8.1.38) is given by formula (8.2.7) with Q given by formula (8.2.21) as follows:

$$u(\hat{x}) = e^{iky} + i\xi g(\hat{x},t_1)|S_1|u_0(t_1), \quad a \to 0. \tag{8.2.22}$$

This formula is valid for $|\hat{x} - t_1| \gg a$, but when $a \to 0$, the formula is valid practically everywhere.

This completes the discussion of wave scattering by one thin cylinder.

If the cylinder is circular, then $|S_1| = 2\pi a$.

8.3 Many-body wave scattering problem equation for the effective field

Consider now the many-body wave scattering problem. In this case,

$$D = \bigcup_{m=1}^{M} D_m, \quad t_m \in D_m, \quad 1 \le m \le M,$$

the distribution law of the domains D_m or, equivalently, the points t_m, is given by formula (8.1.1).

We want to solve problems (8.1.36)–(8.1.38) where now $D = \bigcup_{m=1}^{M} D_m$, the impedances are $\zeta = \zeta_m$,

$$\zeta_m = h(t_m)a^\kappa, \quad 0 \le \kappa < 1, \quad \text{Re} h(\hat{x}) \ge 0, \tag{8.3.1}$$

and $h(t)$ is a continuous function on Ω, which the experimenter can choose as he(she) wishes together with the parameter κ. As $a \to 0$, one has $\zeta \to 0$ and $\xi = \frac{\omega\mu}{\zeta} = \frac{\omega\mu}{h_m}a^{-\kappa}$, where $h_m := h(t_m)$. One has $\xi \to \infty$ as $a \to 0$.

Let us look for the solution of problems (8.1.36)–(8.1.38) of the form

$$u(\hat{x}) = u_0(\hat{x}) + \sum_{m=1}^{M} \int_{S_m} g(\hat{x}, t)\sigma_m(t)dt \sim u_0(\hat{x})$$

$$+ \sum_{m=1}^{M} g(\hat{x}, t_m)Q_m, \quad a \to 0, \tag{8.3.2}$$

where the term

$$J_2 := \sum_{m=1}^{M} \int_{S_m} [g(\hat{x}, t) - g(\hat{x}, t_m)]\,\sigma_m(t)dt \tag{8.3.3}$$

is of higher order of smallness as $a \to 0$ compared with the term

$$J_1 := \sum_{m=1}^{M} g(\hat{x}, t_m)Q_m.$$

Thus, the idea is quite similar to the one used in a study of wave scattering problem for one body.

Let us check that

$$\left| \int_{S_m} [g(\hat{x}, t) - g(\hat{x}, t_m)]\sigma_m(t)dt \right| \ll a\,|\nabla_t g(\hat{x}, t)|_{t=t_m}|\,|Q_m|, \quad a \to 0. \tag{8.3.4}$$

Suppose that $k > 0$ is fixed and $k|\hat{x} - t_m| \gg 1$. Then

$$|g(\hat{x}, t_m)| = O\left(\frac{1}{(k|\hat{x} - t_m|)^{1/2}}\right), \tag{8.3.5}$$

and, if $|t_m - t| \le a$, then

$$|g(\hat{x}, t_m) - g(\hat{x}, t)| = O\left(\frac{ka}{(k|\hat{x} - t_m|)^{1/2}}\right), \tag{8.3.6}$$

so (8.3.4) holds because of the assumption $ka \ll 1$, see (8.2.1).

If $|\hat{x} - t_m| \sim d \gg a$, but $kd \ll 1$, then

$$|g(k|\hat{x} - t|)| = O\left(\ln \frac{1}{kd}\right), \tag{8.3.7}$$

$$|g(\hat{x}, t) - g(\hat{x}, t_m)| = O\left(\frac{a}{d}\right), \tag{8.3.8}$$

because $H_0^1(r) = O\left(\ln\left(\frac{1}{r}\right)\right)$ as $r \to 0$, and $\frac{d}{dr}H_1^0(r) = -H_1^1(r) = O\left(\frac{1}{r}\right)$ as $r \to 0$. Thus, (8.3.4) holds if $a \ll d$. Consequently, formula (8.3.2) holds. This formula reduces the solution to many-body scattering problems (8.1.36)–(8.1.38) to finding numbers Q_m rather than functions $\sigma_m(t)$, $1 \le m \le M$. Let us derive an linear algebraic systems for finding these numbers.

Define the effective field acting on the j-th small body by the formula

$$u_e(\hat{x}) = u_0(\hat{x}) + \sum_{m \neq j} \int_{S_m} g(\hat{x}, t)\sigma_m(t)dt, \tag{8.3.9}$$

and

$$u = u_e + \int_{S_j} g(\hat{x}, t)\sigma_j(t)dt. \tag{8.3.10}$$

Substitute (8.3.10) into the exact boundary condition (8.1.37) and obtain the relation similar to (8.2.13) with u_e replacing u_0. The estimates similar to the ones used in the derivation of the formula (8.2.21) yield a formula similar to (8.2.21), as follows:

$$Q_j = i\xi_j|S_j|u_e(t_j), \quad 1 \le j \le M, \quad \xi = \frac{\omega\mu}{\zeta_j}, \quad \zeta_j = h_j a^\kappa. \tag{8.3.11}$$

Formulas (8.3.9) and (8.3.11) yield the following LAS for the unknown $u_e(t_m)$:

$$u_e(t_j) = u_0(t_j) + i\sum_{m \neq j}^{M} g(t_j, t_m)\xi_m|S_m|u_e(t_m), \quad 1 \le j \le M. \tag{8.3.12}$$

Denote $u_e(t_j) := u_j$, assume for simplicity that the cylinders are identical, denote

$$|S_m| = c_S a,$$

let $g(t_j, t_m) := g_{jm}$, $h_m = \frac{\zeta_m}{a^\kappa}$, and use formula (8.3.1) to rewrite (8.3.12) as

$$u_j = u_{0j} + ic_S \sum_{m\neq j}^{M} g_{jm}\xi'_m u_m a^{1-\kappa}, \quad 1 \leq j \leq M, \quad \xi'_m := \frac{\omega\mu}{h_m}.$$
(8.3.13)

Let us pass to the limit $a \to 0$ in the linear algebraic system (8.3.13). Consider a partition of the domain Ω, where the small cylinders are distributed according to the law (8.1.1), into a union of P small squares Δ_p of the side $b = b(a)$, and assume that

$$a \ll d \ll b,$$

where d is the smallest distance between neighboring cylinders. Within one small square the quantities $g_{jm}\xi'_m u_m$ are equal to $g_{qp}\xi'_p u_p$ up to the terms of higher order of smallness as $a \to 0$. Here $t_p \in \Delta_p$ is an arbitrary point and $t_q \in \Delta_q$, $q \neq p$. Therefore, the sum in (8.3.13) can be written as

$$ic_S \sum_{m\neq j}^{M} g_{jm}\xi'_m u_m a^{1-\kappa} = ic_S \sum_{q\neq p} g_{qp}\xi'_p u_p a^{1-\kappa} \sum_{t_m\in\Delta_p} 1. \quad (8.3.14)$$

By formula (8.1.1), one has

$$a^{1-\kappa} \sum_{t_m\in\Delta_p} 1 = N(t_p)|\Delta_p| := N_p|\Delta_p|, \quad b \to 0, \quad (8.3.15)$$

where $|\Delta_p|$ is the area of Δ_p. Consequently, linear algebraic system (8.3.13) of order M for finding numbers u_m can be reduced to an LAS of much lower order $P \ll M$, namely to the following LAS:

$$u_q = u_{0q} + ic_S \sum_{p\neq q} g_{pq}\xi'_p N_p|\Delta_p|u_p. \quad (8.3.16)$$

This is a discretized version of the following linear integral equation:

$$u(\hat{x}) = u_0(\hat{x}) + ic_S \int_\Omega g(\hat{x},\hat{y})\xi(\hat{y})N(\hat{y})u(\hat{y})d\hat{y} := u_0 + Tu. \quad (8.3.17)$$

Recall that c_S is the constant in the definition $|S_m| = c_S a$, $\xi(\hat{x}) = \frac{\omega\mu}{h(\hat{x})}$, the function $h(\hat{x})$ is defined in formula (8.3.1), the function $N(\hat{y})$ is defined in formula (8.1.1), and $g(\hat{x}, \hat{y}) := \frac{i}{4} H_0^1(k|\hat{x} - \hat{y}|)$, where $\hat{x} = (x_1, x_2)$ and $\hat{y} = (y_1, y_2)$. Do not confuse the function $\xi(\hat{x}) = \frac{\omega\mu}{h(\hat{x})}$ in formula (8.3.17) with ξ in formula (8.1.37).

8.4 Physical properties of the limiting medium

Let us now interpret physically our result. Applying the operator $\nabla^2 + k^2$ to equation (8.3.17) and using the equation $(\nabla^2 + k^2)u_0 = 0$, $(\nabla^2 + k^2)g(\hat{x}, \hat{y}) = -\delta(\hat{x} - \hat{y})$ one obtains

$$\left(\hat{\nabla}^2 + k^2\right) u(\hat{x}) = -i c_S \xi(\hat{x}) N(\hat{x}) u(\hat{x}). \qquad (8.4.1)$$

Rewrite this equation as

$$\left[\hat{\nabla}^2 + k^2 n^2(\hat{x})\right] u(\hat{x}) = 0, \qquad (8.4.2)$$

where

$$n(\hat{x}) := \left[1 + i c_S k^{-2} \xi(\hat{x}) N(\hat{x})\right]^{1/2}, \quad \xi(\hat{x}) = \frac{\omega\mu}{h(\hat{x})}. \qquad (8.4.3)$$

This is a *new refraction coefficient* of the limiting medium obtained by embedding many thin impedance cylinders of an arbitrary cross-section of the length $c_S a$. The functions $N(x) \geq 0$ and $h(\hat{x})$, $\mathrm{Re}\,h(\hat{x}) \geq 0$, are continuous and can be chosen by the experimenter. The condition $\mathrm{Re}\,h(\hat{x}) \geq 0$ guarantees the uniqueness of the solution to problems (8.1.36)–(8.1.38) with

$$\xi(\hat{x}) = \frac{\omega\mu}{h(\hat{x})}.$$

$\mathrm{Re}\,h := h_1 \geq 0$, then $\mathrm{Re}\,\xi \geq 0$. If $\mathrm{Im}\,h := h_2 \geq 0$, then $\mathrm{Im}\,\xi \leq 0$.

By choosing properly $N(\hat{x}) \geq 0$ and $h(\hat{x})$, $\mathrm{Re}\,h(\hat{x}) \geq 0$, one can create a desired refraction coefficient.

Can one create a negative refraction coefficient $n(\hat{x}) < 0$?

Such a question is of interest in the theory of meta-materials.

The square root in formula (8.4.3) one can define as $z^{1/2} := |z|^{1/2} e^{i\frac{\arg z}{2}}$, where $z = |z|e^{i\arg z}$. Let $\theta := \arg z$, then $z^{1/2} = |z|^{1/2} e^{i\frac{\theta}{2}}$.

Thus, $z^{1/2} < 0$ if $\theta = 2\pi$. In our case, $z = 1 + \frac{ic_S k^{-2} N(\hat{x})\omega\mu}{h(\hat{x})}$. If $h = h_1 + ih_2$, where h_1 and h_2 are real numbers, then

$$z = 1 + \frac{c_S k^{-2}\omega\mu N(\hat{x})(ih_1 + h_2)}{h_1^2 + h_2^2} := A + i\eta. \qquad (8.4.4)$$

Suppose that $N(\hat{x})$, $h_1(\hat{x})$, and $h_2(\hat{x})$ are chosen so that

$$A := 1 + \frac{c_S k^{-2}\omega\mu N(\hat{x})h_2(\hat{x})}{h_1^2(\hat{x}) + h_2^2(\hat{x})} > 0, \qquad \eta := \frac{c_S k^{-2}\omega\mu N(\hat{x})h_1(\hat{x})}{h_1^2(\hat{x}) + h_2^2(\hat{x})} < 0.$$
$$(8.4.5)$$

This is possible if, for example, $h_1(\hat{x}) < 0$ and $h_2(\hat{x}) > 0$. Then $z = A + i\eta$, $\eta < 0$, and $\text{Re}z^{1/2}$ is negative while $\text{Im}z^{1/2}$ is positive.

Let us show that such a choice of h_1 is possible to make without violating the uniqueness of the solution of the equation (8.3.17). Let $\zeta = h_1 + ih_2$ and $h_1 < 0$ is sufficiently small. The meaning of the word sufficiently will be clarified in the subsequent argument. Then $z = A + i\eta$, where $\eta < 0$ is small and $A > 0$. For such z one has $z^{1/2} = (A^2 + \eta^2)^{1/2} e^{i\frac{\arg(A+i\eta)}{2}}$, and $\arg(A + i\eta) = 2\pi - \delta$, where $\delta > 0$ is small compared with π. Therefore, $\frac{1}{2}\arg(A + i\eta) = \pi - \frac{\delta}{2} \simeq \pi$. Consequently, $z^{1/2} \simeq Ae^{i\pi} = -A$, so $\text{Re}n(\hat{x}) < 0$ and $\text{Im}n(\hat{x}) > 0$ for this choice of $h(\hat{x})$ and $N(\hat{x})$. If $|A| \gg |\eta|$, then $|\text{Re}n(\hat{x})|$ is much larger than $\text{Im}n(\hat{x})$, and $n(\hat{x})$ is practically negative.

Let us now prove that a sufficiently small negative h_1 still guarantees the uniqueness of the solution to equation (8.3.17).

Lemma 8.4.1. *If h_1 is sufficiently small and negative, then equation (8.3.17) has a unique solution.*

Proof. Equation (8.3.17) is of Fredholm type: since the domain Ω is finite and the kernel of the integral operator T in equation (8.3.17) is weakly singular, the operator T is compact in $L^2(\Omega)$. For $h_1 \geq 0$, this equation is uniquely solvable. Since the set of boundedly invertible linear operators $I-T$ is *open* and the operator $I-T$ is boundedly invertible at $h_1 = 0$, it must be boundedly invertible for sufficiently small negative h_1. Lemma 8.4.1 is proved. $\qquad\square$

8.5   Summary of the results

In this chapter, an analytical solution to an electromagnetic wave scattering problem by many parallel thin impedance cylinders with a crosssection of an arbitrary shape is given.

The existence of the limit of the effective field in the medium where such cylinders are embedded is proved as $a \to 0$ and $M \to \infty$ under a suitable rate. Here a is the characteristic size of the cross-section of a thin cylinder and $M = M(a)$ is the total number of the cylinders. The limiting medium is described by a new refraction coefficient (8.4.3). This coefficient can be created as the experimenter wishes by choosing properly the boundary impedances of the thin cylinders and their distribution function $N(\hat{x})$, see (8.1.1).

Chapter 9

Heat Transfer in a Medium in Which Many Small Bodies are Embedded

9.1 Introduction

In this chapter, heat transfer is considered in a complicated system which consists of many small bodies embedded in a given material and concentrated with a given distribution law in a bounded domain Ω.

Let D_m be a small body, $D = \bigcup_{m=1}^{M} D_m$, $D \subset \Omega \subset \mathbb{R}^3$, S_m is the boundary of D_m, $a = \frac{1}{2}\text{diam}D_m$ is a characteristic size of the small bodies. For simplicity, these bodies are assumed identical, but it is not difficult to generalize the theory to the case when the bodies are not identical. The distribution law for these bodies is given by the formula

$$\mathcal{N}(\Delta) = \frac{1}{a^{2-\kappa}} \int_{\Delta} N(x)dx[1 + o(1)], \quad a \to 0. \qquad (9.1.1)$$

Here $N(x) \geq 0$ is a given continuous function in Ω, $\kappa \in [0, 1)$ is a given number, Δ is an arbitrary open subset of Ω, the boundaries S_m are assumed to be sufficiently smooth, for example, $C^{1,\gamma}$-smooth.

Heat transfer in the system is described by the following equations:

$$u_t = \nabla^2 u + f(x) \quad \text{in } \mathbb{R}^3 \backslash D; \quad u = u(x,t), \quad u_t = \frac{\partial u}{\partial t}, \quad (9.1.2)$$

$$u|_{t=0} = 0, \tag{9.1.3}$$

$$u_N = \zeta_m u \quad \text{on } S_m, \quad 1 \leq m \leq M, \tag{9.1.4}$$

where $f \in L^2(\mathbb{R}^3 \backslash D)$ is compactly supported, ζ_m is a given number,

$$\zeta_m = \frac{h(x_m)}{a^\kappa}, \quad \text{Re} h \geq 0, \quad x_m \in D_m, \tag{9.1.5}$$

and N is the unit normal to S_m pointing out of D_m. The function $h(x)$ in (9.1.5) is a given continuous in Ω function, and κ is a given number.

Our goal is to study the limiting behavior as $a \to 0$ and $M = M(a) \to \infty$ of the heat distribution in Ω, to derive the equation of the limiting distribution of the temperature u, and to give numerical methods for calculating the solution to problems (9.1.2)–(9.1.4) when M is large. We do not assume periodicity in the location of the small bodies and our approach differs from the usual approach developed in theory.

We derive a linear algebraic system (LAS) for finding the Laplace transform of the solution to problems (9.1.2)–(9.1.4) and an integral equation for the limiting temperature as $a \to 0$.

9.2 Derivation of the equation for the limiting temperature

Let us denote by $\mathcal{U} = \mathcal{U}(x, \lambda)$ the Laplace transform of u as follows:

$$\mathcal{U}(x, \lambda) := \int_0^\infty u(x,t) e^{-\lambda t} dt. \tag{9.2.1}$$

Taking the Laplace transform of (9.1.2)–(9.1.4), one gets

$$-\nabla^2 \mathcal{U} + \lambda \mathcal{U} = \lambda^{-1} f(x) \quad \text{in } \Omega, \tag{9.2.2}$$

$$\mathcal{U}_N = \zeta_m \mathcal{U} \quad \text{on } S_m, 1 \leq m \leq M. \tag{9.2.3}$$

Let

$$g := g(x, y) := g(x, y, \lambda) := \frac{e^{-\sqrt{\lambda}|x-y|}}{4\pi|x-y|}, \qquad (9.2.4)$$

$$F(x, \lambda) := \lambda^{-1} \int_\Omega g(x, y) f(y) dy. \qquad (9.2.5)$$

Let us look for the unique solution to problems (9.2.2)–(9.2.3) of the form

$$\mathcal{U} := \mathcal{U}(x, \lambda) = F(x, \lambda) + \sum_{m=1}^{M} \int_{S_m} g(x, s) \sigma_m(s) ds, \qquad (9.2.6)$$

where $\sigma_m(s)$ are unknown functions to be found from the boundary conditions (9.2.3). To use this condition let us define the effective field, acting on m-th particle, by the formula

$$\mathcal{U}_e(x, \lambda) := \mathcal{U}(x, \lambda) - \int_{S_m} g(x, s) \sigma_m(s) ds. \qquad (9.2.7)$$

We define also some operators, as follows:

$$T_m \sigma_m := \int_{S_m} g(s, s') \sigma_m(s') ds', \qquad (9.2.8)$$

$$A_m \sigma_m := 2 \int_{S_m} \frac{\partial g(s, s')}{\partial N_s} \sigma_m(s') ds', \qquad (9.2.9)$$

and set

$$Q_m := \int_{S_m} \sigma_m(s) ds. \qquad (9.2.10)$$

From (9.2.7) and (9.2.3), one obtains

$$\frac{\partial \mathcal{U}_e}{\partial N_s} - \zeta_m \mathcal{U}_e + \frac{A_m \sigma_m - \sigma_m}{2} - \zeta_m T_m \sigma_m = 0, \quad s \in S_m, \qquad (9.2.11)$$

where the known formula was used for the limiting value of the normal derivative of the potential of the single layer when x approaches point $s \in S_m$ from $D'_m := \mathbb{R}^3 \backslash D_m$, as follows:

$$\frac{\partial}{\partial N_s^-} \int_{S_m} g(x, s') \sigma_m(s') ds' = \frac{A_m \sigma_m - \sigma_m}{2}. \qquad (9.2.12)$$

By $\frac{\partial}{\partial N_s^-}$, we denote the above limiting value of the normal derivative on S.

We do not want to solve the system of boundary integral equations (9.2.11), $1 \leq m \leq M$, for σ_m. If M is large, this is practically hardly feasible. We want to reduce the problem to finding numbers Q_m (see (9.2.10)) rather than unknown functions $\sigma_m(t)$. To do this, rewrite formula (9.2.6) as

$$\mathcal{U} = F(x, \lambda) + \sum_{m=1}^{M} g(x, x_m) Q_m$$

$$+ \sum_{m=1}^{M} \int_{S_m} [g(x, s) - g(x, x_m)] \sigma_m(s) ds, \qquad (9.2.13)$$

and prove that if the bodies D_m are sufficiently small and the smallest distance $d = d(a)$ between neighboring bodies satisfies the condition

$$\lim_{a \to 0} \frac{a}{d(a)} = 0, \qquad (9.2.14)$$

or, practically,

$$a \ll d, \qquad (9.2.15)$$

then

$$\left| \int_{S_m} [g(x, s) - g(x, s_m)] \sigma_m(s) ds \right| \ll |g(x, x_m) Q_m|, \quad 1 \leq m \leq M. \qquad (9.2.16)$$

This relation is proved in Section 2.1. It allows one to calculate U by the following formula (cf (9.2.13)):

$$\mathcal{U} = F(x, \lambda) + \sum_{m=1}^{M} g(x, x_m) Q_m, \quad a \to 0, \qquad (9.2.17)$$

and, therefore, reduce the problem to finding numbers Q_m.

To find these numbers, let us integrate the exact boundary condition (9.2.11) over S_m and neglect terms of the higher order of smallness as $a \to 0$. Let us estimate the order of these terms.

The first term is

$$\int_{S_m} \frac{\partial \mathcal{U}_e}{\partial N_s} ds = \int_{D_m} \nabla^2 \mathcal{U}_e dx = O\left(a^3\right). \tag{9.2.18}$$

The second term is

$$-\zeta_m \int_{S_m} \mathcal{U}_e(s, \lambda) ds = -\zeta_m \mathcal{U}_e(x_m, \lambda)|S_m| = -c_S h_m \mathcal{U}_e(x_m, \lambda) a^{2-\kappa}, \tag{9.2.19}$$

where we have used the relation (9.1.5), set $|S_m| = c_S a^2$, $c_S = \text{const}$, and denote by $|S_m|$ the surface area of S_m.

The third term is

$$\int_{S_m} \frac{A_m \sigma_m - \sigma_m}{2} ds = -Q_m. \tag{9.2.20}$$

Here, as in Section 1.1, the relation

$$\int_{S_m} ds A_m \sigma_m = -\int_{S_m} \sigma_m(t) dt, \quad a \to 0, \tag{9.2.21}$$

was used.

The fourth term is

$$-\zeta_m \int_{S_m} ds \int_{S_m} g(s, s') \sigma_m(s') ds' = -\int_{S_m} ds' \sigma_m(s') \int_{S_m} ds g(s, s') \zeta_m$$

$$= o(Q_m), \quad a \to 0, \tag{9.2.22}$$

provided that

$$\zeta_m a = o(1), \quad a \to 0, \tag{9.2.23}$$

which is the case because of (9.1.5) and the assumption $\kappa < 1$. From (9.2.18)–(9.2.23), it follows that

$$Q_m = -c_S h_m \mathcal{U}_e(x_m, \lambda) a^{2-\kappa}, \quad 1 \le m \le M. \tag{9.2.24}$$

Denote $\mathcal{U}_e(x_m, \lambda) := \mathcal{U}_m$, $F_m := F(x_m, \lambda)$, and write

$$\mathcal{U}_e(x, \lambda) = F(x, \lambda) + \sum_{m \neq j}^{M} g(x, x_m) Q_m. \qquad (9.2.25)$$

Substitute into (9.2.25) x_j in place of x, use formula (9.2.24), and get the following LAS for the unknown numbers $\mathcal{U}_m$:

$$\mathcal{U}_j = F_j - c_S \sum_{m \neq j}^{M} g_{jm} h_m \mathcal{U}_m a^{2-\kappa}, \quad 1 \leq j \leq M. \qquad (9.2.26)$$

Here

$$g_{jm} := g(x_j, x_m), \quad F_j = F(x_j, \lambda). \qquad (9.2.27)$$

If the LAS (9.2.26) is solved numerically for $\mathcal{U}_m$, $1 \leq m \leq M$, then formulas (9.2.25) and (9.2.24) allow one to calculate the effective field $\mathcal{U}_e$ in Ω.

Let us reduce the order of the LAS (9.2.26) and derive the integral equation for the limiting effective field $\mathcal{U} = \lim_{a \to 0} \mathcal{U}_e$ in Ω.

Consider a partition of Ω into a union of $P \ll M$ small cubes Δ_p with the size $b \gg d \gg a$, $b = b(a)$,

$$\lim_{a \to 0} \frac{d}{b} = 0. \qquad (9.2.28)$$

For all $x_m \in \Delta_p$, one has

$$a^{2-\kappa} \sum_{x_m \in \Delta_p} g_{qm} h_m \mathcal{U}_m = g_{qp} h_p \mathcal{U}_p a^{2-\kappa} \sum_{x_m \in \Delta_p} 1, \quad q \neq p, \qquad (9.2.29)$$

because in a small cube Δ_p all the terms under the first sum are equal up to the terms of the higher order of smallness and they are equal to the value of these terms at a fixed point $x_p \in \Delta_p$. The sum on the right-hand side of formula (9.2.29) multiplied by $a^{2-\kappa}$ can be calculated by the formula (9.1.1) as follows:

$$a^{2-\kappa} \sum_{x_m \in \Delta_p} 1 = a^{2-\kappa} \mathcal{N}(\Delta_p) = N(x_p)|\Delta_p|, \quad a \to 0. \qquad (9.2.30)$$

Therefore, formula (9.2.26) can be written as

$$\mathcal{U}_q = F_q - c_S \sum_{p \neq q} g_{qp} h_p \mathcal{U}_p N_p |\Delta_p|, \quad 1 \leq q \leq P, \tag{9.2.31}$$

where $N_p := N(x_p)$. This is a linear algebraic system of a reduced order $P \ll M$. It can be considered as a collocation method for numerical solution of the following integral equation for the limiting field $\mathcal{U}$ in Ω:

$$\mathcal{U}(x, \lambda) = F(x, \lambda) - c_S \int_\Omega g(x, y) h(y) N(y) \mathcal{U}(y) dy. \tag{9.2.32}$$

Convergence of the collocation method for solving equation (9.2.32) is proved in Chapter 12.

Let us summarize the result we have obtained.

Theorem 9.2.1. *The limit $\mathcal{U}(x, \lambda)$ of the field $\mathcal{U}_e(x, \lambda)$ as $a \to 0$ does exist and solves equation (9.2.32). The solution to this equation is unique.*

Proof. Only the last statement of Theorem 9.2.1 is not yet proved. To prove it, it is sufficient to prove that the corresponding homogeneous equation (with $F = 0$) has only the trivial solution. Apply the operator $-\nabla^2 + \lambda$ to (9.2.32) with $F = 0$ and get

$$\left(-\nabla^2 + \lambda + c_S h(x) N(x)\right) \mathcal{U} = 0 \quad \text{in } \mathbb{R}^3. \tag{9.2.33}$$

This is a Schrödinger equation with a potential $q(x) = c_S N(x) h(x)$ which has compact support Ω, and $\mathcal{U}$ decays at infinity exponentially as follows from the homogeneous equation (9.2.32). The assumption Re$h \geq 0$ (see (9.1.5)) guarantees that the only solution to equation (9.2.33) is the trivial solution $\mathcal{U} = 0$. Let us prove this claim.

Multiply (9.2.33) by $\overline{\mathcal{U}}$ and integrate the result over $D_R = \{x : |x| \leq R\}$. The result is

$$\int_{D_R} |\nabla \mathcal{U}|^2 dx + \lambda \int_{D_R} |\mathcal{U}|^2 dx + \int_{D_R} q(x) |\mathcal{U}|^2 dx = 0. \tag{9.2.34}$$

Since $\lambda \geq 0$, the assumption Re$q \geq 0$ and formula (9.2.34) imply $\mathcal{U} = 0$. The assumption Re$q \geq 0$ is equivalent to the assumption Re$h \geq 0$, which was done in formula (9.1.5).

Theorem 9.2.1 is proved. $\qquad\qquad\qquad\qquad\qquad\qquad \square$

9.3 Various results

Applying the operator $-\nabla^2 + \lambda$ to equation (9.2.32), one gets

$$(-\nabla^2 + \lambda + q(x))\mathcal{U} = \frac{f(x)}{\lambda} \quad \text{in } \mathbb{R}^3; \quad q(x) := c_S N(x) h(x). \quad (9.3.1)$$

Take the inverse Laplace transform of (9.3.1) and obtain

$$u_t = \nabla^2 u - q(x)u + f(x) \quad \text{in } \mathbb{R}^3, \quad\quad (9.3.2)$$

$$u|_{t=0} = 0, \quad\quad (9.3.3)$$

where condition (9.3.3) follows from the formula

$$u(0) = \lim_{\lambda \to \infty} (\lambda \mathcal{U}), \quad\quad (9.3.4)$$

and the exponential decay of $\mathcal{U}$ as $\lambda \to \infty$.

Suppose one is interested in the average temperature

$$\lim_{T \to \infty} \frac{1}{T} \int_0^T u(x,t)dt = \lim_{\lambda \to 0} \lambda \mathcal{U}(x,\lambda) := \psi(x). \quad\quad (9.3.5)$$

Existence of this limit is proved as follows. The function $\psi(x)$ can be calculated by the formula

$$\psi(x) = (I + B)^{-1}\phi, \quad\quad (9.3.6)$$

where

$$\phi := \int_{\mathbb{R}^3} \frac{f(y)dy}{4\pi|x-y|}, \quad B\psi := \int_{\mathbb{R}^3} \frac{q(y)\psi(y)dy}{4\pi|x-y|}. \quad\quad (9.3.7)$$

Let us prove the above claim.

Lemma 9.3.1. *Formula* (9.3.4) *holds.*

Proof. Assume that $u_t \in L^1(0,\infty)$. Then one has

$$\bar{u}(\lambda) := \mathcal{U}(\lambda) := \int_0^\infty e^{-\lambda t}u(t)dt = \frac{e^{-\lambda t}u(t)}{-\lambda}\Big|_0^\infty + \frac{1}{\lambda}\int_0^\infty e^{-\lambda t}u_t(t)dt.$$
$$(9.3.8)$$

Multiply both sides by λ and then take $\lambda \to \infty$ and get (9.3.4).
Lemma 9.3.1 is proved.   □

Lemma 9.3.2. *Formula* (9.3.5) *holds.*

Proof. One has

$$\int_\lambda^\infty \frac{\overline{u}(\sigma)}{\sigma} d\sigma = \frac{1}{t} \int_0^t u(s) ds, \qquad (9.3.9)$$

where the overbar denotes the Laplace transform.

Let $v = v(t)$ be some locally integrable function. If $\lim_{t\to\infty} v(t) = 0$, then

$$\lim_{\lambda\to 0} \lambda \int_0^\infty e^{-\lambda t} v(t) dt = 0. \qquad (9.3.10)$$

Indeed, $|v(t)| \leq \epsilon$ if $t > N(\epsilon)$. One has

$$\left| \lambda \int_0^{N(\epsilon)} e^{-\lambda t} v(t) dt \right| \leq \lambda \int_0^{N(\epsilon)} |v(t)| dt < \epsilon, \qquad (9.3.11)$$

if $0 < \lambda < \lambda(\epsilon)$, where $\epsilon > 0$ is an arbitrary small fixed number. Also

$$\left| \lambda \int_{N(\epsilon)}^\infty e^{-\lambda t} v(t) dt \right| \leq \epsilon\lambda \int_{N(\epsilon)}^\infty e^{-\lambda t} dt = \epsilon e^{-\lambda N(\epsilon)} \leq \epsilon. \qquad (9.3.12)$$

Thus, the conclusion (9.3.10) follows.

If $\lim_{t\to\infty} v(t) = v_\infty$, then

$$\lim_{\lambda\to 0} \lambda \int_0^\infty e^{-\lambda t} v(t) dt = v_\infty, \qquad (9.3.13)$$

because (9.3.10) implies

$$\lim_{\lambda\to 0} \lambda \int_0^\infty e^{-\lambda t} (v(t) - v_\infty) dt = 0 \qquad (9.3.14)$$

and

$$\lambda \int_0^\infty e^{-\lambda t} v_\infty dt = v_\infty \lambda \int_0^\infty e^{-\lambda t} dt = v_\infty. \qquad (9.3.15)$$

Formulas (9.3.9) and (9.3.13) imply

$$\lim_{t\to\infty} \frac{1}{t} \int_0^t u(s) ds = \lim_{\lambda\to 0} \lambda \int_\lambda^\infty \frac{\overline{u}(\sigma)}{\sigma} d\sigma = \lim_{\lambda\to 0} \lambda\overline{u}(\lambda). \qquad (9.3.16)$$

The last equality follows from the relation

$$\lim_{\lambda \to 0} \int_\lambda^\infty \frac{\lambda}{\sigma^2} f(\sigma) d\sigma = f(0), \tag{9.3.17}$$

which is valid if $\lim_{\sigma \to 0} f(\sigma) = f(0)$ exists. In our case, $f(\sigma) = \sigma \overline{u}(\sigma) = \sigma \mathcal{U}(\sigma)$, so formula (9.3.5) is obtained. Lemma 9.3.2 is proved. $\square$

Lemma 9.3.3. *Formula* (9.3.6) *holds.*

Proof. Let us apply the operator $(-\nabla^2)^{-1}$ with the kernel $\frac{1}{4\pi|x-y|}$ to equation (9.3.1) and get

$$\mathcal{U} + \lambda \int_{\mathbb{R}^3} \frac{u(y,\lambda)dy}{4\pi|x-y|} + \int_{\mathbb{R}^3} \frac{q(y)\mathcal{U}(y,\lambda)dy}{4\pi|x-y|} = \frac{1}{\lambda} \int_{\mathbb{R}^3} \frac{f(y)dy}{4\pi|x-y|}. \tag{9.3.18}$$

Multiply both sides by λ, then take $\lambda \to 0$, and obtain

$$\psi + B\psi = \int_{\mathbb{R}^3} \frac{f(y)dy}{4\pi|x-y|} := \phi, \tag{9.3.19}$$

because

$$\lim_{\lambda \to 0} \lambda^2 \int_{\mathbb{R}^3} \frac{\mathcal{U}(y,\lambda)dy}{4\pi|x-y|} = 0, \tag{9.3.20}$$

and

$$\lim_{\lambda \to 0} \lambda \int_{\mathbb{R}^3} \frac{q(y)\mathcal{U}(y,\lambda)dy}{4\pi|x-y|} = \int_{\mathbb{R}^3} \frac{q(y)\psi(y)}{4\pi|x-y|} dy := B\psi. \tag{9.3.21}$$

From (9.3.19), the conclusion of Lemma 9.3.3 follows. $\square$

Finally, let us discuss the *invertibility* of the operator $I + B$.

Since B is compact in $L^2(\Omega)$, and support of q is a compact set $\overline{\Omega}$, the operator $(I + B)^{-1}$ exists and is bounded if and only if the equation

$$u + Bu = 0 \tag{9.3.22}$$

has only the trivial solution. Applying to this equation the operator $-\nabla^2$ one gets

$$\left[-\nabla^2 + q(x)\right] u = 0 \quad \text{in } \mathbb{R}^3. \tag{9.3.23}$$

Thus, the invertibility of the operator $I+B$ is reduced to the existence of a nontrivial solution to equation (9.3.23) in the class of functions decaying at infinity as $O\left(\frac{1}{|x|}\right)$. If $\mathrm{Re}q \geq 0$, then multiplying equation (9.3.23) by $\bar{u}$, where the bar denotes complex conjugate, integrate over the ball $|x| \leq R$, then integrate by parts and get

$$\int_{|x| \leq R} (|\nabla u|^2 + q|u|^2)dx - \int_{|S|=R} \bar{u}u_N ds = 0. \qquad (9.3.24)$$

Since $q = 0$ for $|x| \geq R_0$, where $\Omega \subset B_{R_0}$, $u = O\left(\frac{1}{|x|}\right)$, $u_N = O\left(\frac{1}{|x|^2}\right)$, so

$$\lim_{R \to \infty} \int_{|s|=R} \bar{u}u_N ds = 0. \qquad (9.3.25)$$

Taking the real part of equation (9.3.24), one concludes that $u = 0$ if $\mathrm{Re}q \geq 0$. We have proved the following lemma.

Lemma 9.3.4. *If $supp(q) \subset \Omega$, and Ω is a bounded region, and if $\mathrm{Re}q \geq 0$, then the operator $(I + B)^{-1}$ is boundedly invertible in $L^2(B_{R_0})$.*

9.4 Summary of the results

In this chapter, the basic result is formulated in Theorem 9.2.1. Other results are formulated in Section 9.3.

Chapter 10

Quantum-Mechanical Wave Scattering by Many Potentials with Small Support

10.1 Problem formulation

In this chapter, quantum-mechanical wave scattering by many potentials with small support is studied. The number of the potentials tends to infinity while the characteristic radius a of their supports tends to zero. The limiting medium is described by a refraction coefficient that can be of desired form if the potentials with small support are chosen properly.

Let us assume that a bounded domain $\Omega \subset \mathbb{R}^3$ is filled with a material with a known refraction coefficient $n_0(x)$, $\mathrm{Im}n_0^2(x) \geq 0$,

$$n_0(x) = 1 \quad \text{in } \Omega' := \mathbb{R}^3\backslash\Omega, \quad \mathrm{Im}n_0^2(x) \geq 0. \qquad (10.1.1)$$

The wave scattering problem by this material consists of finding the solution to the equations

$$L_0u_0 := \left[\nabla^2 + k^2n_0^2(x)\right]u_0 = 0 \quad \text{in } \mathbb{R}^3, \quad k = \text{const} > 0, \qquad (10.1.2)$$

$$u_0 = e^{ik\alpha \cdot x} + v, \qquad (10.1.3)$$

where $\alpha \in S^2$ is a given unit vector and v is the scattered field satisfying the radiation condition

$$\frac{\partial v}{\partial r} - ikv = o\left(\frac{1}{r}\right), \quad r := |x| \to \infty. \qquad (10.1.4)$$

Lemma 10.1.1. *Under the above assumptions, the function u_0 exists and is unique.*

Proof.   Let us write equation (10.1.2) as

$$L_0 u_0 = \left(\nabla^2 + k^2 - q_0(x) \right) u_0 = 0, \qquad (10.1.5)$$

where

$$q_0(x) := q_0(x, k) = k^2 - k^2 n_0^2(x), \qquad (10.1.6)$$

$$q_0(x) = 0, \quad x \in \Omega'. \qquad (10.1.7)$$

Let $G(x, y)$ solve the equation

$$L_0 G(x, y) = -\delta(x - y) \quad \text{in } \mathbb{R}^3. \qquad (10.1.8)$$

Denote

$$g(x, y) = \frac{e^{ik|x-y|}}{4\pi|x-y|}. \qquad (10.1.9)$$

Then

$$\left(\nabla^2 + k^2 \right) g(x, y) = -\delta(x - y) \quad \text{in } \mathbb{R}^3. \qquad (10.1.10)$$

The function u_0 solves the equation

$$u_0 = e^{ik\alpha \cdot x} - \int_\Omega g(x, y) q_0(y) u_0(y) dy := e^{ik\alpha \cdot x} - T u_0. \qquad (10.1.11)$$

This is a Fredholm-type equation. Existence and uniqueness of its solution u_0 will be proved if one proves that the corresponding homogeneous equation has only the trivial solution. If v solves the homogeneous equation

$$v = -Tv, \qquad (10.1.12)$$

then

$$\left(\nabla^2 + k^2 - q_0(x) \right) v = 0, \qquad (10.1.13)$$

and v satisfies condition (10.1.4). Multiply equation (10.1.13) by $\bar{v}$, the bar stands for complex conjugate, the complex conjugate of equation (10.1.13) by u, subtract from the first equation the second, integrate over the ball $|x| \le R$, apply Green's formula, and take $R \to \infty$.

The result is

$$\lim_{R\to\infty}\left(\int_{|s|=R}\left(\overline{v}\frac{\partial v}{\partial N}-v\frac{\partial\overline{v}}{\partial N}\right)ds-\int_{|x|\leq R}[q_0(x)-\overline{q}_0(x)]\,|v|^2dx\right)=0.$$
(10.1.14)

The radiation condition (10.1.4) implies

$$\lim_{R\to\infty}\left(2ik\int_{|s|=R}|v|^2ds-2i\int_{|x|\leq R}\mathrm{Im}q_0(x)|v|^2dx\right)=0.\quad(10.1.15)$$

The assumption $\mathrm{Im}n_0^2(x)\geq 0$ is equivalent to $\mathrm{Im}q_0(x)\leq 0$. Therefore, it follows from (10.1.15) that

$$\lim_{R\to\infty}\int_{|s|=R}|v|^2ds=0.\tag{10.1.16}$$

Any solution to equation (10.1.5) with a compactly supported potential q_0, such that condition (10.1.16) holds, is equal to zero identically (see Ramm (1986), p. 25). Lemma 10.1.1 is proved. $\square$

Let us assume that M potentials $q_m(x)=n_m\chi_m(x)$ are embedded in the domain Ω, where $n_m=\mathrm{const}$, $\chi_m(x)$ is the characteristic function of the domain D_m, $\chi_m(x)=1$ in D_m, 0 in $D'_m:=\mathbb{R}^3\backslash D_m$. The volume V_a of the domains D_m is assumed independent of m, a point $x_m\in D_m$, $V_a=c_Va^3$, the constant c_V characterizes the shape of D_m. If D_m are balls, then $c_V=\frac{4\pi}{3}$.

Let $\nu(x)$ be a continuous in Ω function such that

$$\nu(x_m)=n_m^2.$$

We assume that the domains D_m are distributed in Ω according to the following law:

$$\mathcal{N}(\Delta):=\sum_{x_m\in\Delta}1=\frac{1}{V_a}\int_\Delta N(x)dx[1+o(1)],\quad a\to 0,\quad(10.1.17)$$

where $\Delta\subset\Omega$ is an arbitrary subdomain of Ω, $\mathcal{N}(\Delta)$ is the number of potentials of small support in Δ, $N(x)\geq 0$ is a given continuous function that can be chosen by an experimenter.

The scattering problem by many potentials with small support is described by the equation

$$\left(L_0 + k^2 \sum_{m=1}^{M} n_m \chi_m(x)\right) u = 0 \quad \text{in } \mathbb{R}^3, \tag{10.1.18}$$

$$u = u_0 + V, \tag{10.1.19}$$

where V is the scattered field satisfying the radiation condition (10.1.4) and u_0 is the solution to problems (10.1.2)–(10.1.3), corresponding to the absence of potentials with small support. The solution to problems (10.1.18)–(10.1.19) exists, is unique, and depends on the parameter a, the characteristic size of the small support, $u = u_a$. We will prove that the limit

$$\lim_{a \to 0} u_a(x) = u_e(x) \tag{10.1.20}$$

exists and satisfies the following integral equation:

$$u_e(x) = u_0(x) + k^2 \int_{\Omega} G(x, y)\nu(y)N(y)u_e(y)dy. \tag{10.1.21}$$

Theorem 10.1.1. *If the potentials of small support are distributed in Ω according to the law (10.1.17), then the limit (10.1.20) exists in $C(\Omega)$ and equation (10.1.21) holds.*

A proof of this theorem is given in the next section.

Let us draw some physical conclusions from equation (10.1.21). Applying the operator L_0 to this equation and using equation (10.1.8), one gets

$$L_0 u_e = -k^2 \nu(x) N(x) u_e(x). \tag{10.1.22}$$

This equation can be written as

$$\left(\nabla^2 + k^2 (n_0^2(x) + \nu(x)N(x))\right) u_e(x) = 0. \tag{10.1.23}$$

Therefore, the new medium, obtained after the embedding of many potentials with small support, has a new refraction coefficient

$$n(x) := \left(n_0^2(x) + \nu(x)N(x)\right)^{1/2}. \tag{10.1.24}$$

Since the function $\nu(x)$ is at the disposal of the experimenter, one can choose it so that $n(x)$ is a desirable function. The function $N(x) \geq 0$ can be fixed *a priori*, for example, one can choose $N(x) = \text{const} > 0$, and then $\nu(x)$ can be chosen so that the function (10.1.24) is a desired function. For example, if one wishes to have negative $n(x)$, then one should have $\text{Re}(n_0^2(x) + \nu(x)N(x)) > 0$, and $\text{Im}(n_0^2(x) + \nu(x)N(x))$ negative and very small. In this case, the argument of the quantity $n_0^2(x) + \nu(x)N(x)$ will be close to 2π and $n(x)$ will be close to a negative number. We use the formula $z^{1/2} = |z|^{1/2}e^{i\varphi/2}$, where z is a complex number, $z^{1/2}$ is its square root, $|z|$ is the absolute value of z, and φ is its argument.

The uniqueness lemma, Lemma 10.1.1, was proved under the assumption $\text{Im}q_0(x) \leq 0$. However, the integral equation (10.1.11), on which the proof was based, is an equation of Fredholm-type. If it is uniquely solvable for some q_0, it will be uniquely solvable for q sufficiently close to q_0, in particular for $q = q_0 + i\epsilon$, where $\epsilon > 0$ is sufficiently small. If $\text{Im}q_0 = 0$, then $\text{Im}(q_0 + i\epsilon) > 0$ and still equation (10.1.11), or a similar equation, corresponding to problems (10.1.18)–(10.1.19),

$$u(x) = u_0(x) + k^2 \int_\Omega G(x,y) \sum_{m=1}^{M} n_m \chi_m(y) u(y) dy, \qquad (10.1.25)$$

will be uniquely solvable, the limiting equation (10.1.21) will be uniquely solvable, and formula (10.1.24) yields $n(x)$ with $\text{Re}n(x) < 0$ and $\text{Im}n(x) \approx 0$.

10.2   Proofs

In this section, Theorem 10.1.1 is proved. The proof uses some lemmas.

Lemma 10.2.1. *One has*

$$\lim_{|x-y| \to 0} |x - y| G(x,y) = \frac{1}{4\pi}, \qquad (10.2.1)$$

$$\sup_{x,y \in \mathbb{R}^3} \left(|x - y| G(x,y) \right) \leq c. \qquad (10.2.2)$$

Proof of Lemma 10.2.1. The integral equation for G is

$$G(x,y) = g(x,y) - \int_\Omega g(x,z)q_0(z)G(z,y)dz. \qquad (10.2.3)$$

Let us consider as the new unknown function

$$G(x,y) - g(x,y) := h(x,y). \qquad (10.2.4)$$

Then

$$h(x,y) = -h_0(x,y) - \int_\Omega g(x,z)q_0(z)h(z,y)dz := -h_0 - Th, \qquad (10.2.5)$$

where

$$h_0(x,y) := \int_\Omega g(x,z)q_0(z)g(z,y)dz. \qquad (10.2.6)$$

We assume that

$$\sup_{z\in\Omega} |q_0(z)| \le c, \qquad (10.2.7)$$

so that

$$|h_0(x,y)| \le c \int_\Omega \frac{dz}{|x-z||z-y|} \le c(\Omega), \qquad (10.2.8)$$

where by c we denote various constants. The function $h_0(x,y)$ is continuous in $\Omega \times \Omega$, and equation (10.2.5) is considered in $C(\Omega)$ for any fixed $y \in \Omega$. Equation (10.2.5) is of Fredholm type. It is uniquely solvable for any $h_0 \in C(\Omega)$ if and only if the homogeneous equation (10.2.5) has only the trivial solution. Denote by $\psi(x)$ the solution to the homogeneous equation (10.2.5). Then, applying the operator $\nabla^2 + k^2$ to this equation, one gets

$$\left[\nabla^2 + k^2 - q_0\right]\psi = 0 \quad \text{in } \mathbb{R}^3, \qquad (10.2.9)$$

and ψ satisfies the radiation condition. By Lemma 10.1.1, $\psi = 0$. Therefore, equation (10.2.5) is uniquely solvable and the operator

$(I + T)^{-1}$ is bounded in $C(\Omega)$. Consequently,

$$\max_{x \in \Omega} |h(x, y)| \leq c \max_{x \in \Omega} |h_0(x, y)|. \qquad (10.2.10)$$

The right-hand side of (10.2.10) does not depend on y. Therefore,

$$\max_{x,y \in \Omega} |h(x, y)| \leq c. \qquad (10.2.11)$$

Consequently,

$$\max_{x,y \in \Omega} |G(x, y) - g(x, y)| \leq c. \qquad (10.2.12)$$

One can extend the unique solution to equation (10.2.3) in Ω to the unique solution in $\mathbb{R}^3$ because the function $g(x, y)$ and the integral in (10.2.3) are defined for all $x \in \mathbb{R}^3$. Consequently, estimate (10.2.12) extends to

$$\max_{x,y \in \mathbb{R}^3} |G(x, y) - g(x, y)| \leq c. \qquad (10.2.13)$$

Multiplying (10.2.13) by $|x - y|$ and taking $|x - y| \to 0$, one gets (10.2.1).

If y runs in a bounded domain, then it follows from equation (10.2.3) that

$$|G(x, y)| \leq \frac{c(R)}{1 + |x|}, \quad |y| \leq R, \quad x \in \mathbb{R}^3. \qquad (10.2.14)$$

Since $G(x, y) = G(y, x)$, a similar estimate holds for y, as follows:

$$|G(x, y)| \leq \frac{c(R)}{1 + |y|}, \quad |x| \leq R, \quad y \in \mathbb{R}^3. \qquad (10.2.15)$$

Thus, from equation (10.2.3), boundedness of Ω, and estimates (10.2.14)–(10.2.15), it follows that

$$|G(x, y)| \leq \frac{c}{|x - y|} + \frac{c}{|x||y|}, \quad |x| \geq R, \quad |y| \geq R. \qquad (10.2.16)$$

Finally,

$$\frac{|x-y|}{|x||y|} \le \frac{|x|+|y|}{|x||y|} \le \frac{1}{|y|} + \frac{1}{|x|} \le \frac{2}{R}, \quad |x| \ge R, \quad |y| \ge R. \tag{10.2.17}$$

From estimates (10.2.16) and (10.2.17), one gets estimate (10.2.2). Lemma 10.2.1 is proved. □

Proof of Theorem 10.1.1. Let us write equation (10.1.8) as the following integral equation:

$$u = u_0 + k^2 \sum_{m=1}^{M} g(x,y)\nu(x_m)u(y)dy. \tag{10.2.18}$$

This equation can be discretized as follows:

$$u(x_j) = u_0(x_j) + k^2 \sum_{\substack{m \ne j}}^{M} g(x_j, x_m)\nu(x_m)u(x_m)|D_m|, \quad 1 \le j \le M, \tag{10.2.19}$$

where $|D_m| = V_a$ is the volume of D_m.

By the distribution law (10.1.17), one has the following in every small subdomain Δ_p:

$$V_a \sum_{x_m \in \Delta_p} 1 = N(x_p)|\Delta_p|(1 + o(1)), \quad \text{diam}\Delta_p \to 0, \tag{10.2.20}$$

where $x_p \in \Delta_p$ is an arbitrary point and $|\Delta_p|$ is the volume of Δ_p. Consider a partition of Ω into a union of small cubes Δ_p, $1 \le p \le P$, with a side $b = b(a)$,

$$\lim_{a \to 0} \frac{a}{b(a)} = 0, \tag{10.2.21}$$

and rewrite the sum in (10.2.19) as following:

$$k^2 \sum_{\substack{m \ne j}}^{M} g(x_j, x_m)\nu(x_m)u(x_m)|D_m|$$

$$= k^2 \sum_{\substack{p \ne q}}^{P} g(x_q, x_p)\nu(x_p)N(x_p)u(x_p)|\Delta_p|. \tag{10.2.22}$$

Then formula (10.2.19) yields

$$u(x_q) = u_0(x_q) + k^2 \sum_{p \neq q}^{P} g(x_q, x_p)\nu(x_p)N(x_p)u(x_p)|\Delta_p|. \quad (10.2.23)$$

This is a collocation method for solving the integral equation (10.1.21). Convergence of this collocation method is proved in Chapter 12. This implies both conclusions of Theorem 10.1.1.

Theorem 10.1.1 is proved. □

10.3   Summary of the results

It is proved in this chapter that one can distribute many potentials of small support in a given bounded domain so that the limiting medium will have a desired refraction coefficient given by formula (10.1.24). The results are formulated in Theorem 10.1.1.

Chapter 11

Some Results from the Potential Theory

11.1 Potentials of the simple and double layers

Let $D \subset \mathbb{R}^3$ be a bounded domain with a boundary $S \in C^{1,\gamma}$. This means that in local coordinates the equation of the surface is $z = f(x,y)$, where z-axis is directed along the normal to S which points out of D and x, y are axes in the tangent plane to S at the origin, which is located on S. The function $f(x,y)$ is continuously differentiable and

$$\left|\nabla f(x,y) - \nabla f(x',y')\right| \le c \left(\left|x-x'\right|^2 + \left|y-y'\right|^2\right)^{\gamma/2},$$

$$\gamma = \text{const}, \quad 0 < \gamma \le 1. \tag{11.1.1}$$

By $c > 0$, various constants are denoted.

Define potentials of the single and double layers as follows:

$$u(x) := \int_S g(x,t)\sigma(t)dt; \quad w(x) := \int_S \frac{\partial g(x,t)}{\partial N_t}\mu(t)dt. \tag{11.1.2}$$

The functions σ and μ are assumed Hölder-continuous, that is,

$$\left|\sigma(t) - \sigma(t')\right| \le c|t-t'|^\nu, \quad \left|\mu(t) - \mu(t')\right| \le c|t-t'|^\nu;$$

$$t, t' \in S, \quad 0 < \nu \le 1. \tag{11.1.3}$$

The function $g(x,t)$ is

$$g(x,t) = \frac{e^{ik|x-t|}}{4\pi|x-t|}. \tag{11.1.4}$$

Theorem 11.1.1. *The function $u(x)$ is continuous in $\mathbb{R}^3$. It is infinitely differentiable inside D and outside D, but its normal derivative has a jump across S, namely*

$$u_N^{\pm} = \frac{A\sigma \pm \sigma}{2}, \quad A\sigma := 2\int_S \frac{\partial g(s,t)}{\partial N_s}\sigma(t)dt. \tag{11.1.5}$$

The sign u_N^+ (u_N^-) denotes limiting value of the derivative $\nabla_x u \cdot N_s$ when the point $x \to s \in S$ along the normal N_s, $x \in D$ $(x \in D' := \mathbb{R}^3 \backslash D)$.

Theorem 11.1.2. *The function w is infinitely differentiable in D and in D' and has a jump across S:*

$$w^{\pm} = \frac{A'\mu \mp \mu}{2}, \quad A'\mu := 2\int_S \frac{\partial g(s,t)}{\partial N_t}\mu(t)dt. \tag{11.1.6}$$

The results of these theorems are classical and can be found, for example, in Günter (1967). For completeness, we give their proofs. These proofs differ from the proofs in Günter (1967). We need some preparations.

Lemma 11.1.1. *One has*

$$\eta(x) := \int_S \frac{\partial g_0(x,t)}{\partial N_t}dt = \begin{cases} -1, & x \in D, \\ -\dfrac{1}{2}, & x \in S, \\ 0, & x \in D', \end{cases} \quad g_0(x,t) := \frac{1}{4\pi|x-t|}.$$

$$\tag{11.1.7}$$

Proof. If $x \in D'$, then applying Green's formula one gets

$$\int_S \frac{\partial g_0(x, t)}{\partial N_t} dt = \int_D \nabla_y^2 g_0(x, y) dy = -\int_D \delta(x - y) dy = 0, \quad x \in D'.$$

$$(11.1.8)$$

Similarly,

$$\int_S \frac{\partial g_0(x, t)}{\partial N_t} dt = -\int_D \delta(x - y) dy = -1, \quad x \in D. \qquad (11.1.9)$$

Finally, if $s \in S$, one has

$$\int_S \frac{\partial g_0(s, t)}{\partial N_t} dt = \lim_{\delta \to 0} \int_{|s-t| \geq \delta} \frac{\partial g_0(s, t)}{\partial N_t} dt = -\frac{1}{2}. \qquad (11.1.10)$$

To prove (11.1.10), take a semisphere K_δ centered at s, of radius $\delta \to 0$ and belonging to D'. Denote the union of this semisphere K_δ and the part of S described by the inequality $|s - t| \geq \delta$ by S_δ. Since S_δ contains s inside, one has by formula (11.1.9) that

$$\int_{S_\delta} \frac{\partial g_0(s, t)}{\partial N_t} dt = -1. \qquad (11.1.11)$$

Therefore,

$$\lim_{\delta \to 0} \int_{|s-t| \geq \delta} \frac{\partial g_0(x, t)}{\partial N_t} dt = -1 - \lim_{\delta \to 0} \int_{K_\delta} \frac{\partial g_0(x, t)}{\partial N_t} dt := -1 - I,$$

$$(11.1.12)$$

where I is the integral over K_δ. One has

$$\lim_{\delta \to 0} \int_{K_\delta} \frac{\partial}{\partial N_t} \frac{1}{4\pi |s - t|} dt = -\lim_{\delta \to 0} \frac{1}{4\pi \delta^2} \int_{K_\delta} dt = -\frac{1}{2}. \qquad (11.1.13)$$

From (11.1.10) and (11.1.13), the middle formula (11.1.7) follows. Lemma 11.1.1 is proved. $\qquad \square$

From Lemma 11.1.1, it follows that

$$\int_S A_0 \sigma ds = -\int_S \sigma ds.$$

The function $\frac{\partial g(x,t)}{\partial N_t} - \frac{\partial g_0(x,t)}{\partial N_t}$ is continuous when x crosses S along the normal. Therefore, formulas (11.1.5) and (11.1.6) follow from similar formulas, in which g is replaced by g_0.

Proof of Theorem 11.1.2. Let us prove formula (11.1.6) with g_0 in place of g and A_0' in place of A'. One has

$$I := \int_S \frac{\partial g_0(x,t)}{\partial N_t} \mu(t) dt$$

$$= \int_S \frac{\partial g_0(x,t)}{\partial N_t} \mu(s) dt + \int_S \frac{\partial g_0(x,t)}{\partial N_t} \left[\mu(t) - \mu(s) \right] dt$$

$$:= I_1 + I_2. \tag{11.1.14}$$

From (11.1.7), it follows that

$$I = \eta(x)\mu(s) + I_2. \tag{11.1.15}$$

Let us show that I_2 is continuous when x crosses S along the normal. If this is proved, then formula (11.1.15) shows that

$$I^+ = -\mu + I_2, \quad I^- = I_2, \quad I_0 = -\frac{\mu}{2} + I_2, \tag{11.1.16}$$

where I_0 is the value of I on S. Thus,

$$I^\pm = I_0 \mp \frac{\mu(s)}{2}, \quad I_0(s) = \frac{A'\mu}{2} = \int_S \frac{\partial g_0(s,t)}{\partial N_t} \mu(t) dt. \tag{11.1.17}$$

Thus, the proof is completed if one proves that I_2 is continuous when x crosses S along N_s.

One has

$$\left| \frac{\partial g_0(x,t)}{\partial N_t} [\mu(t) - \mu(s)] \right| \le c|t - s|^\nu \frac{|(t-x)^o \cdot N_t|}{|x-t|^2},$$

$$(t-x)^o := \frac{t-x}{|t-x|}. \tag{11.1.18}$$

Let us use the local coordinates with the origin at the point s, $s = (0,0,0)$ where the z-axis is directed along N_s, and $x = (0,0,x)$, $x \ge 0$. We denote by the same letter the vector $(0,0,x)$ and its z-coordinate x. This does not bring any confusion and simplifies the notations. In these coordinates, one has

$$t_3 = f(t_1, t_2), \quad f \le 0, \quad x - f(t_1, t_2) \ge x,$$

$$|x-t|^2 = |x - f(t_1, t_2)|^2 + |t_1|^2 + |t_2|^2;$$

$$|t-s|^2 = |f(t_1, t_2)|^2 + |t_1|^2 + |t_2|^2. \tag{11.1.19}$$

Therefore,

$$\frac{|t-s|^{\nu}}{|x-t|^2} \le \frac{\left(|f(t_1,t_2)|^2 + |t_1|^2 + |t_2|^2\right)^{\frac{\nu}{2}}}{|x-f(t_1,t_2)|^2 + |t_1|^2 + |t_2|^2}$$

$$\le \frac{1}{\left(|x-f(t_1,t_2)|^2 + |t_1|^2 + |t_2|^2\right)^{1-\frac{\nu}{2}}}, \qquad (11.1.20)$$

where the inequalities $x \ge 0$ and $f \le 0$ were used, so $x - f(t_1,t_2) \ge |f(t_1,t_2)|$.

Formula (11.1.20) shows that the function on the left-hand side of formula (11.1.18) is absolutely integrable with respect to (t_1,t_2) uniformly with respect to $x \ge 0$.

In the aforementioned argument, we did not use the assumption $f \in C^{1,\gamma}$, only the existence of the normal N_s was used. However, if $x \le 0$, that is, the point $x \in D$, and if vector x tends to s along N_s, then we have to use the assumption $f \in C^{1,\gamma}$. Namely, in our local coordinates $f(s) = f(0,0) = 0$, $\nabla f(0,0) = 0$, and, by formula (11.1.1) with $(x',y') = (0,0)$, one has

$$|f(t_1,t_2)| = |f(t_1,t_2) - f(0,0)|$$

$$= \left|\int_0^1 \nabla f(pt_1, pt_2) \cdot t\, dp\right| \le c\left(|t_1|^2 + |t_2|^2\right)^{\frac{1+\gamma}{2}}, \quad (11.1.21)$$

where $t = (t_1, t_2)$.

This estimate yields the following (see (11.1.19)):

$$\frac{|t-s|^{\nu}}{|x-t|^2} \le \frac{\left(|f(t_1,t_2)|^2 + |t_1|^2 + |t_2|^2\right)^{\frac{\nu}{2}}}{|x-f(t_1,t_2)|^2 + |t_1|^2 + |t_2|^2}$$

$$\le c\frac{(|t_1|^2 + |t_2|^2)^{\nu/2}}{|t_1|^2 + |t_2|^2} \le c\left(|t_1|^2 + |t_2|^2\right)^{2-\frac{\nu}{2}}. \qquad (11.1.22)$$

Estimates (11.1.20) for $x \ge 0$ and (11.1.22) for $x < 0$ show that the integrand in I_2, namely, the function (11.1.18), is absolutely integrable with respect to (t_1,t_2) uniformly with respect to $x \in N_s$. This implies that I_2 is continuous when x crosses S along the normal. Theorem 11.1.2 is proved. $\qquad\square$

Proof of Theorem 11.1.1. This proof is based on the following idea. Suppose that we have proved that if $x < 0$, then the function

$$\phi(x) := \int_S \frac{\partial g(x,t)}{\partial N_s} \sigma(t)dt + \int_S \frac{\partial g(x,t)}{\partial N_t} \sigma(t)dt \qquad (11.1.23)$$

is continuous when x crosses S along N_s. Then Theorem 11.1.1 is proved: its conclusion follows from Theorem 11.1.2 and from the continuity of $\phi(x)$, because the jump of the normal derivative of the single layer potential differs by sign from the jump of the double layer potential. The formula for the jump of the double layer potential is already proved in Theorem 11.1.2. The expression

$$\frac{\partial g(x,t)}{\partial N_s} := \nabla_x g(x,t) \cdot N_s \qquad (11.1.24)$$

allows one to write formula (11.1.23) as

$$\begin{aligned}
\phi(x) &:= \int_S \left(\nabla_x g(x,t) \cdot N_s + \nabla_t g(x,t) \cdot N_t \right) dt \\
&= \int_S \left(\nabla_x g(x,t) + \nabla_t g(x,t) \right) \cdot N_s dt + \int_S \nabla_t g(x,t) \cdot (N_t - N_s) dt \\
&:= J_1 + J_2.
\end{aligned} \qquad (11.1.25)$$

Let us check that J_1 and J_2 are continuous functions of x when x crosses S along N_s. From our assumption about S, it follows that

$$|N_s - N_t| \le c|s - t|^\gamma, \quad \gamma > 0. \qquad (11.1.26)$$

Therefore, the continuity of J_2 is proved as we have proved the continuity of I_2 defined in (11.1.14). Let us prove the continuity of J_1. One has

$$\nabla_x g(x,t) + \nabla_t g(x,t) = \nabla_x g(x,t) - \nabla_x g(x,t) = 0. \qquad (11.1.27)$$

Thus, $\phi(x)$ is continuous when x crosses S along N_s. Consequently, formula (11.1.23) implies

$$\phi(s) = u_N^+ + w^+ = u_N^- + w^- = u_0 + w_0, \qquad (11.1.28)$$

where

$$u_N^{\pm} = \lim_{\epsilon \to \pm 0} \int_S \frac{\partial g(s \mp \epsilon N_s)}{\partial N_s} \sigma(t)dt, \quad u_0 := \int_S \frac{\partial g(s, t)}{\partial N_s} \sigma(t)dt,$$

$$(11.1.29)$$

$$w^{\pm} = w_0 \mp \frac{\sigma}{2}, \quad w_0 := \frac{A'\sigma}{2}. \tag{11.1.30}$$

Formula (11.1.30) is actually the formula (11.1.6) with σ in place of μ.

From (11.1.28)–(11.1.30), one gets

$$u_N^+ = u_0 + \frac{\sigma}{2}, \quad u_N^- = u_0 - \frac{\sigma}{2}, \quad u_0 = \frac{A\sigma}{2}. \tag{11.1.31}$$

Theorem 11.1.1 is proved. □

One can also prove (see Günter (1967)) that

$$w_N^+ = w_N^- \tag{11.1.32}$$

provided that the density μ of w is continuously differentiable.

Lemma 11.1.2. *If $S \in C^2$ and $\mu \in C^1$, then formula (11.1.32) holds in the sense*

$$\lim_{0 < h \to 0} \left(\frac{\partial w(s + hN_s)}{\partial N_s} - \frac{\partial w(s - hN_s)}{\partial N_s} \right) = 0, \quad \forall s \in S. \tag{11.1.33}$$

Proof. It is sufficient to prove formula (11.1.33) for w_0, where

$$w_0(x) := \int_S \frac{\partial g_0(x, t)}{\partial N_t} \mu(t)dt, \quad g_0(x, t) = \frac{1}{4\pi|x - t|} = \frac{1}{4\pi r_{xt}},$$

$$(11.1.34)$$

because $g - g_0$ is sufficiently smooth as $r_{xt} \to 0$.

Furthermore, without loss of generality one may assume that $\mu(s) = 0$, because

$$w_0(x, \mu) = w_0(x, \mu(s)) + w_0(x, \mu(t) - \mu(s)), \tag{11.1.35}$$

and

$$w_0(x, \mu(s)) = \mu(s)\eta(x), \quad \eta(x) = \begin{cases} -1, & x \in D, \\ 0, & x \in D'. \end{cases} \tag{11.1.36}$$

Thus, for $w_0(x, \mu(s))$ formula (11.1.33) holds, and

$$|\mu(t) - \mu(s)| \le c|t - s|, \tag{11.1.37}$$

since μ is continuously differentiable. Let us prove formula (11.1.33) for w_0. Without loss of generality one may assume that $\mu(s) = 0$. Clearly,

$$\frac{\partial g_0(s,t)}{\partial N_t} = \frac{N_t \cdot (s - t)}{4\pi r_{st}^3}, \tag{11.1.38}$$

where the dot stands for the inner product of two vectors. Since $S \in C^2$ and the angles between N_t and $\frac{s-t}{|s-t|}$ and N_s and $\frac{s-t}{|s-t|}$ tend to $\pi/2$ as $t \to s$, the following inequalities hold:

$$|N_t \cdot (s - t)| \le c|s - t|^2, \quad |N_s \cdot (s - t)| \le c|s - t|^2. \tag{11.1.39}$$

Here and in the following, $c > 0$ stands for various constants independent of s and t. One has

$$\frac{\partial}{\partial N_s} \frac{\partial g_0(s,t)}{\partial N_t} = \frac{N_t \cdot N_s}{4\pi r_{st}^3} - \frac{3N_t \cdot (s - t)N_s \cdot (s - t)}{4\pi r_{st}^5}. \tag{11.1.40}$$

Using inequality (11.1.39), one gets

$$\frac{|N_t \cdot (s - t)N_s \cdot (s - t)|}{r_{st}^5} \le \frac{c}{r_{st}}. \tag{11.1.41}$$

Consequently,

$$-\frac{3}{4\pi} \lim_{0<h\to 0} \int_S \left(\frac{3N_t \cdot (s + hN_s - t)N_s \cdot (s + hN_s - t)}{r_{(s+hN_s)t}^5} - \right.$$
$$\left. \frac{3N_t \cdot (s - hN_s - t)N_s \cdot (s - hN_s - t)}{r_{(s-hN_s)t}^5} \right) \mu(t)dt = 0. \tag{11.1.42}$$

Consider the first term in (11.1.40).

We want to prove the following relation:

$$\lim_{0<h\to 0}\int_S \left(\frac{1}{|s+hN_s-t|^3}-\frac{1}{|s-hN_s-t|^3}\right)\mu(t)dt=0, \quad (11.1.43)$$

assuming that $\mu(s)=0$, that is, $|\mu(t)|\le c|t|^\lambda$ for $|t|\le\epsilon$. We take $\lambda=1$ for simplicity.

One has

$$I:=\frac{1}{|s-t+hN_s|^3}-\frac{1}{|s-t-hN_s|^3}$$

$$=\frac{|s-t-hN_s|^3-|s-t+hN_s|^3}{|s-t+hN_s|^3|s-t-hN_s|^3}$$

$$=\frac{(|s-t-hN_s|^2+|s-t+hN_s|^2+|s-t-hN_s||s-t+hN_s|)}{|s-t+hN_s|^3|s-t-hN_s|^3(|s-t+hN_s|+|s-t-hN_s|)}$$

$$\cdot (|s-t-hN_s|^2-|s-t+hN_s|^2), \quad (11.1.44)$$

$$|s-t-hN_s|^2-|s-t+hN_s|^2=-4h(s-t)\cdot N_s. \quad (11.1.45)$$

Each of the terms in (11.1.44) is treated similarly. Consider, for example, the term

$$I_1:=-4h\int_S\frac{(s-t)\cdot N_s|s-t-hN_s|^2\mu(t)dt}{|s-t+hN_s|^3|s-t-hN_s|^3(|s-t+hN_s|+|s-t-hN_s|)}. \quad (11.1.46)$$

One has

$$|I_1|\le ch\int_S\frac{|s-t|^2|\mu(t)|dt}{|s-t+hN_s|^3|s-t-hN_s|(|s-t+hN_s|+|s-t-hN_s|)}$$

$$\le ch\int_S\frac{|\mu(t)|dt}{|s-t+hN_s|^3}. \quad (11.1.47)$$

Choose the coordinate system with the origin at the point s, x_3-axis along N_s, and x_1,x_2 plane tangent to S at the point s, and use the estimate

$$|\mu(t)|\le c|t|, \quad |t|<\epsilon, \quad (11.1.48)$$

where $\epsilon > 0$ is a fixed small number, to get

$$|I_1| \le ch \int_0^\epsilon \frac{drr^2}{[r^2 + |h - t_3|^2]^{3/2}} + O(h). \tag{11.1.49}$$

The term $O(h)$ comes from the part of the integral in (11.1.46) taken over $|t| > \epsilon$.

In our coordinate system, $t_3 = O(r^2)$. We have to prove that

$$\lim_{h \to 0} h \int_0^\epsilon \frac{drr^2}{[r^2 + |h - t_3|^2]^{3/2}} := \lim_{h \to 0} hI_2 = 0. \tag{11.1.50}$$

One has, for sufficiently small h, for example, for $h < 1/4$, the following inequality:

$$hI_2 \le ch \int_0^\epsilon \frac{drr^2}{[r^2 + h^2]^{3/2}}. \tag{11.1.51}$$

Let $u := r^2$. Then, integrating by parts, one gets

$$hI_2 \le ch \int_0^{\epsilon^2} \frac{duu^{1/2}}{[u + h^2]^{3/2}}$$

$$= ch \left(-2[u + h^2]^{-1/2} u^{1/2} \Big|_0^{\epsilon^2} + \int_0^{\epsilon^2} \frac{du}{\sqrt{u(u + h^2)}} \right). \tag{11.1.52}$$

The term

$$h \frac{\epsilon}{\sqrt{\epsilon^2 + h^2}} \le h \tag{11.1.53}$$

tends to zero as $h \to 0$.

The other term also tends to zero:

$$\lim_{h \to 0} h \int_0^{\epsilon^2} \frac{du}{\sqrt{u(u + h^2)}} = 0. \tag{11.1.54}$$

To check (11.1.54), one calculates the following integral:

$$\int_0^{\epsilon} \frac{du}{\sqrt{u^2 + h^2 u}} = \ln\left(u + \frac{h^2}{2} + \sqrt{\left(u + \frac{h^2}{2} \right)^2 - \frac{h^4}{4}} \right) \Bigg|_0^{\epsilon^2}$$

$$= \ln \frac{\epsilon^2 + \frac{h^2}{2} + \sqrt{\left(\epsilon^2 + \frac{h^2}{2} \right)^2 - \frac{h^4}{4}}}{\frac{h^2}{2}}$$

$$\leq \ln \frac{\epsilon^2 + \frac{h^2}{2}}{\frac{h^2}{2}}. \tag{11.1.55}$$

It is not difficult now to check that

$$\lim_{h \to 0} h \ln \frac{\epsilon^2 + \frac{h^2}{2}}{\frac{h^2}{2}} = 0 \tag{11.1.56}$$

for any fixed $\epsilon > 0$.

In a similar way the other terms in (11.1.44) are treated.
Lemma 11.1.2 is proved. □

In Lemma 11.1.2, it was assumed that the points $s + hN$ and $s - hN$ are at equal distance h from S. This assumption can be dropped. In Günter (1967), pp. 73–76, formula (11.1.31) is proved without this assumption, but the proof is longer and more complicated that the one given above.

Lemma 11.1.3. *If J is a tangential to S continuous vector field and $S \in C^{1,\lambda}, 0 < \lambda \leq 1$, then the following formula is valid:*

$$\lim_{x \to s^-} \left[N_s, \nabla \times \int_S g(x,t) J(t) dt \right] = \int_S [N_s, [\nabla_s g(s,t), J(t)]] dt + \frac{J(s)}{2}. \tag{11.1.57}$$

Proof. One has

$$\left[N_s, \nabla \times \int_S g(x,t) J(t) dt \right] = \int_S [N_s, [\nabla_x g(x,t), J(t)]] dt$$

$$= \int_S \left(\nabla_x g(x,t) N_s \cdot J(t) - J(t) \frac{\partial g(x,t)}{\partial N_s} \right) dt$$

$$:= b_1 + b_2. \tag{11.1.58}$$

Since $J(t)$ is tangential to S, one has

$$|N_s \cdot J(t)| \le c|t - s|^\lambda, \quad |\nabla_s g(s,t) N_s \cdot J(t)| \le c|s - t|^{-2+\lambda}.$$
$$(11.1.59)$$

Consequently, the integral b_1 is continuous when x crosses S at the point s along N_s. The integral b_2 is analogous to the minus normal derivative of the single-layer potential. Thus, formula (11.1.57) follows.

Lemma 11.1.3 is proved.   □

One can prove the following formula:

$$\nabla_x \int_S g(x,t)\sigma(t)dt \bigg|_{x \to s^-} = \int_S \nabla_s g(s,t)\sigma(t)dt - \frac{\sigma(t)}{2}N_s. \quad (11.1.60)$$

The proof is similar to the proof of formula (11.1.5).

11.2 Replacement of the surface potentials

The aim of this section is to show that potential of single layer $u(\sigma)$ can be replaced by a potential of double layer $w(\mu)$ in the bounded connected smooth domain D, and vice versa.

In the complement of D, that is, in the region $D' := \mathbb{R}^3 \backslash D$, we give a necessary and sufficient condition for $w(x, \mu) = u(x, \sigma)$, $\forall x \in D'$. In other words, we give a necessary and sufficient condition for a single-layer potential to be equal to a double-layer potential, and vice versa. These conditions allow one to replace single-layer potential by a double-layer one, and vice versa.

Let us start with some preparations.

Let

$$(\nabla^2 + k^2)\psi = 0, \quad \text{in } D, \quad \psi|_S = f \qquad (11.2.1)$$

and

$$(\nabla^2 + k^2)\varphi = 0, \quad \text{in } D, \quad \varphi_N|_S = f. \qquad (11.2.2)$$

The corresponding homogeneous problems, that is, problems with $f = 0$, let us denote $(11.2.1_0)$ and $(11.2.2_0)$. If M is a subspace, let us write $\dim M = 0$ if and only if $M = \{0\}$.

Lemma 11.2.1. *If problem* $(11.2.1_0)$ *has only the trivial solution, then* $\dim N(Q) = 0$ *and* $\dim N(I - A) = 0$. *Conversely, if* $\dim N(Q) = 0$ *or* $\dim N(I - A) = 0$, *then problem* $(11.2.1_0)$ *has only the trivial solution.*

Recall that

$$A\sigma := 2 \int_S \frac{\partial g(s,t)}{\partial N_s} \sigma(t)dt, \quad A'\mu := 2 \int_S \frac{\partial g(s,t)}{\partial N_t} \mu(t)dt, \quad (11.2.3)$$

and

$$Q\sigma := \int_S g(s,t)\sigma(t)dt. \quad (11.2.4)$$

Proof of Lemma 11.2.1.

Let us assume that $\dim N(I - A) = 0$, and prove that problem $(11.2.1_0)$ has only the trivial solution.

One has, using the Fredholm alternative and compactness of A in $L^2(S)$,

$$0 = \dim N(I - A) = \dim N(I - A^*) = \dim N(I - \bar{A}')$$
$$= \dim N(I - A'). \quad (11.2.5)$$

Here A^* is the adjoint to A operator in $L^2(S)$, the bar stands for complex conjugate, and A' is the operator with the transposed kernel $A(t,s)$.

If problem $(11.2.1_0)$ has a solution u, then, by Green's formula,

$$u(x) = \int_S \left(g(x,t)u_N(t) - g_N(x,t)u(t)\right)dt$$
$$= \int_S g(x,t)\sigma(t)dt, \quad \sigma := u_N. \quad (11.2.6)$$

Since $u|_S = 0$, the function u solves the problem

$$(\nabla^2 + k^2)u = 0, \quad \text{in } D', \quad u|_S = 0, \quad (11.2.7)$$

and satisfies the radiation condition

$$u_r - iku = o\left(\frac{1}{r}\right), \quad r = |x| \to \infty. \quad (11.2.8)$$

Thus,

$$u(x) = 0, \quad \text{in } D'. \tag{11.2.9}$$

Consequently, $u_N^- = 0$, and $u_N^- = \frac{1}{2}(A - I)\sigma = 0$. Therefore, our assumption $\dim N(I - A) = 0$ implies $\sigma = 0$. Thus, $u = 0$ in D. So, problem (11.2.1$_0$) has only the trivial solution.

Let us now assume that $\dim N(Q) = 0$ and prove that problem (11.2.1$_0$) has only the trivial solution. As before, if u solves problem (11.2.1$_0$), then it can be represented by formula (11.2.6). Thus, $Q\sigma = 0$, and by our assumption, $\sigma = 0$. Therefore, $u = 0$ in D, so problem (11.2.1$_0$) has only the trivial solution.

Let us assume that problem (11.2.1$_0$) has only the trivial solution and prove that $\dim N(Q) = 0$ and $\dim N(I - A) = 0$.

Suppose that $Q\sigma = 0$ and let

$$u := \int_S g(x, t)\sigma(t)dt.$$

If $Q\sigma = 0$, then

$$u = 0, \quad \text{on } S, \tag{11.2.10}$$

and

$$(\nabla^2 + k^2)u = 0, \quad \text{in } D. \tag{11.2.11}$$

By our assumption, the solution in D to the problems (11.2.10)–(11.2.11), which is problem (11.2.1$_0$), is $u = 0$ in D.

The function $u = 0$ in D' also. Indeed, u solves the problem

$$(\nabla^2 + k^2)u = 0 \quad \text{in } D', \quad u|_S = 0, \tag{11.2.12}$$

and satisfies the radiation condition. Thus, $u = 0$ in D'.

Since $u = 0$ in D' and in D, one gets

$$\sigma = u_N^+ - u_N^- = 0. \tag{11.2.13}$$

Therefore, $\dim N(Q) = 0$.

Now, let us prove that $\dim N(I - A) = 0$. Define $u(x)$ by formula (11.2.6). Then u solves problem (11.2.7) and satisfies the radiation condition (11.2.8). This implies that $u = 0$ in D'.

Furthermore, u solves the problem

$$(\nabla^2 + k^2)u = 0 \quad \text{in } D, \quad u|_S = 0. \tag{11.2.14}$$

By our assumption, $u = 0$ in D. So, $u = 0$ in D' and in D. Consequently, $\sigma = 0$ and

$$\dim N(I - A) = 0. \tag{11.2.15}$$

Lemma 11.2.1 is proved. □

Lemma 11.2.2. *If problem* $(11.2.2_0)$ *has only the trivial solution, then* $\dim N(I + A') = 0$. *Conversely, if* $\dim N(I + A') = 0$, *then problem* $(11.2.2_0)$ *has only the trivial solution.*

Proof of Lemma 11.2.2.

Let $A'\mu + \mu = 0$ and assume that $\mu \neq 0$. Consider $w(\mu)$. One has

$$w^- = \frac{A'\mu + \mu}{2} = 0. \tag{11.2.16}$$

Since w solves problem (11.2.7) and satisfies the radiation condition, it follows that

$$w = 0, \quad \text{in } D'. \tag{11.2.17}$$

Thus,

$$w_N^- = 0, \quad \text{on } S. \tag{11.2.18}$$

By the relation (11.1.32), it follows from (11.2.18) that

$$w_N^+ = 0, \quad \text{on } S. \tag{11.2.19}$$

Therefore, w solves problem $(11.2.2_0)$. By our assumption, $w = 0$ in D. Therefore,

$$A'\mu - \mu = 0. \tag{11.2.20}$$

From (11.2.20) and (11.2.16), it follows that

$$\mu = 0. \tag{11.2.21}$$

Thus, $\dim N(I + A') = 0$, as claimed.

Conversely, if $\mu + A'\mu = 0, \mu \neq 0$, then $w = w(x, \mu)$ has the property

$$w^- = 0 \text{ on } S. \qquad (11.2.22)$$

This, equation (11.2.7) for w and the radiation condition (11.2.8) for w imply

$$w = 0, \quad \text{in } D'. \qquad (11.2.23)$$

Thus, $w_N^- = 0$ and, by formula (11.1.32), one gets

$$w_N^+ = 0. \qquad (11.2.24)$$

Consequently, w solves equation (11.2.11) and satisfies boundary condition (11.2.24).

If $\mu \neq 0$, then $w \not\equiv 0$ in D, because otherwise $w^+ = 0$ and $\mu = w^+ - w^- = 0$, which is a contradiction. Consequently, if $\dim N(I + A') = 0$, then problem (11.2.2$_0$) has only the trivial solution. Lemma (11.2.2) is proved.   □

Denote by m, respectively, m', the dimension of the subspace consisting of the solution to problem (11.2.1$_0$), respectively, (11.2.2$_0$).

Lemma 11.2.3. *One has* $\dim N(I + A') = m'$ *and* $\dim N(I - A) = m$.

Proof.   Let $\psi \neq 0$ solve problem (11.2.1$_0$).

As in the proof of Lemma 11.2.1, one obtains formula (11.2.6) and checks that $\sigma := \psi_N^+$ solves the equation $(I - A)\sigma = 0$. In other words, $\sigma \in N(I - A)$.

Thus, there is a correspondence between solutions to problem (11.2.1$_0$) and elements of the set $N(I - A)$.

Let us verify that this correspondence preserves linear independence of the solutions. Suppose that $\{\psi_j\}_{j=1}^m$ is a linearly independent system of the solutions to problem (11.2.1$_0$),

$$\psi := \sum_{j=1}^m c_j \psi_j(x), \qquad (11.2.25)$$

where c_j are constants.

Define

$$\sigma := \psi_N^+ := \sum_{j=1}^m c_j \psi_{jN}^+ := \sum_{j=1}^m c_j \sigma_j. \tag{11.2.26}$$

Then

$$(I - A)\sigma = 0. \tag{11.2.27}$$

We have assumed that

$$\left\{ \sum_{j=1}^m c_j \psi_j(x) = 0 \quad \text{in } D \right\} \Rightarrow \forall c_j = 0. \tag{11.2.28}$$

To each $\sigma_j = \psi_{jN}^+$, there corresponds $\psi_j(x)$ by formula $\psi_j(x) = \int_S g(x,t)\sigma_j(t)dt$. If $\sum_{j=1}^m c_j \sigma_j = 0$, then $\sum_{j=1}^m c_j \psi_j(x) = 0$ in D. Therefore, relations (11.2.26) and (11.2.28) imply

$$\sum_{j=1}^m c_j \sigma_j = 0 \quad \Rightarrow \quad \forall c_j = 0. \tag{11.2.29}$$

This means that if $\{\psi_j\}_{j=1}^m$ are linearly independent, so are $\{\sigma_j\}_{j=1}^m$. Let us establish the converse.
Suppose $(I - A)\sigma_j = 0$ and

$$\sum_{j=1}^m c_j \sigma_j = 0 \quad \Rightarrow \quad \forall c_j = 0. \tag{11.2.30}$$

We want to construct a solution ψ_j to problem (11.2.1$_0$), corresponding to σ_j, and check that the system $\{\psi_j\}_{j=1}^m$ is linearly independent. Given σ_j, let $\psi_j := \psi(x, \sigma_j)$ be defined by the formula

$$\psi_j(x) = \int_S g(x,t)\sigma_j(t)dt. \tag{11.2.31}$$

Since $(I - A)\sigma_j = 0$, it follows that

$$\psi_{jN}^- = 0. \tag{11.2.32}$$

Consequently, as we have argued in detail previously,

$$\psi_j(x) = 0 \quad \text{in } D', \tag{11.2.33}$$

so

$$\psi_j^- = 0 \quad \text{on } S. \tag{11.2.34}$$

Since $\psi^- = \psi^+$, one has

$$\psi_j^+ = 0 \quad \text{on } S. \tag{11.2.35}$$

Thus, ψ_j solves problem (11.2.1$_0$).

Moreover, the function σ_j in (11.2.31) can be expressed as

$$\sigma_j = \psi_{jN}^+ - \psi_{jN}^- = \psi_{jN}^+. \tag{11.2.36}$$

Consequently, if (11.2.30) holds, then

$$\sum_{j=1}^{m} c_j \psi_{jN}^+ = 0 \quad \Rightarrow \quad \forall c_j = 0. \tag{11.2.37}$$

Therefore, if

$$\sum_{j=1}^{m} c_j \psi_j(x) = 0 \quad \text{in } D, \tag{11.2.38}$$

then

$$\sum_{j=1}^{m} c_j \psi_{jN}^+ = 0, \tag{11.2.39}$$

and by (11.2.37), it follows that $\forall c_j = 0$. In this argument, m is the dimension of the subspace of the solution to problem (11.2.1$_0$).

Suppose now that φ is a nontrivial solution to problem (11.2.2$_0$). An application of Green's formula yields

$$\varphi(x) = -\int_S \frac{\partial g(x,t)}{\partial N_t} \varphi^+(t) dt = w(x, -\varphi^+). \tag{11.2.40}$$

Thus,

$$\mu := -\varphi^+ \quad \text{on } S, \quad \mu \not\equiv 0. \tag{11.2.41}$$

By the assumption

$$\varphi_N^+ = 0, \tag{11.2.42}$$

formula (11.1.32) yields

$$\varphi_N^- = 0. \tag{11.2.43}$$

This, as above, allows one to conclude that

$$\varphi(x) = 0 \quad \text{in } D'. \tag{11.2.44}$$

Consequently,

$$\varphi^- = 0. \tag{11.2.45}$$

Therefore, by formula (11.1.6), one gets

$$\mu + A'\mu = 0. \tag{11.2.46}$$

Thus, to every nontrivial solution to problem $(11.2.2_0)$ there corresponds a nontrivial solution to equation (11.2.46).

Let us check that this correspondence preserves linear independence of the solutions.

If a set $\{\varphi_j\}_{j=1}^{m'}$ of solutions to problem $(11.2.2_0)$ is linearly independent, that is

$$\sum_{j=1}^{m'} c_j \varphi_j(x) = 0 \quad \text{in } D \quad \Rightarrow \quad \forall c_j = 0, \tag{11.2.47}$$

then

$$\sum_{j=1}^{m'} c_j \varphi_j^+ = 0 \quad \Rightarrow \quad \forall c_j = 0. \tag{11.2.48}$$

Indeed, if $\sum_{j=1}^{m'} c_j \varphi_j^+ = 0$, then

$$0 = w\left(x, -\sum_{j=1}^{m'} c_j \varphi_j^+\right) = \sum_{j=1}^{m'} c_j w(x, \varphi_j^+) = \sum_{j=1}^{m'} c_j \varphi_j(x). \tag{11.2.49}$$

This and (11.2.47) imply that $\forall c_j = 0$.

Consequently, if a set $\{\mu_j\}_{j=1}^{m'}$ of solutions to equation (11.2.46) is linearly independent, that is,

$$\sum_{j=1}^{m'} c_j \mu_j = 0 \quad \Rightarrow \quad \forall c_j = 0, \tag{11.2.50}$$

then, consider the functions

$$w_j = w(x, \mu_j), \quad 1 \le j \le m', \tag{11.2.51}$$

and note

$$\sum_{j=1}^{m'} c_j w(x, \mu_j) = w\left(x, \sum_{j=1}^{m'} c_j \mu_j\right) = 0 \quad \text{in } D. \tag{11.2.52}$$

If (11.2.50) holds, then

$$\sum_{j=1}^{m'} c_j w(x, \mu_j) = 0 \quad \text{in } D \quad \Rightarrow \quad \forall c_j = 0, \tag{11.2.53}$$

because

$$w(x) = \sum_{j=1}^{m} c_j w(x, \mu_j) = 0 \quad \text{in } D \tag{11.2.54}$$

implies $w_N^- = w_N^+ = 0$.

If $w_N^- = 0$, then using the arguments given above, one gets $w = 0$ in D'. Therefore, (11.2.54) implies

$$\sum_{j=1}^{m'} c_j \mu_j = 0. \tag{11.2.55}$$

Therefore, $\forall c_j = 0$.

Lemma 11.2.3 is proved. □

Let us prove our first theorem.

Theorem 11.2.1. *For every $w(\mu)$ in a bounded domain D there exists a unique $u(\sigma)$ such that $w(\mu) = u(\sigma)$ in D.*

Conversely, for every $u(\sigma)$ there exists a unique $w(\mu)$ such that $u(\sigma) = w(\mu)$ in D.

Proof. If $w(\mu)$ is given, then

$$w^+ = (A'\mu - \mu)/2 = u(\sigma) = Q\sigma \quad \text{on } S. \tag{11.2.56}$$

Consider this equation as an equation for σ. If problem $(11.2.1_0)$ has only the trivial solution, then $N(Q) = \{0\}$ by Lemma 11.2.1. Therefore, equation (11.2.56) has no more than one solution.

To prove the existence of the solution to equation (11.2.56) provided that problem $(11.2.1_0)$ has only the trivial solution, one uses the fact that $Q : H^0 \to H^1$ is an isomorphism if $N(Q) = \{0\}$ and $Q : H^0 \to H^1$ is a Fredholm operator, see Lemma 11.2.4 that follows. Here $H^l = H^l(S)$ is the Sobolev space $W^{l,2}(S)$ and S is a smooth surface.

If $u = u(x, \sigma)$ is given, then equation (11.2.56) is a Fredholm-type equation for μ. This equation is uniquely solvable since the corresponding homogeneous equation has only the trivial solution. Indeed, if $\mu \neq 0$ solves the equation $A'\mu - \mu = 0$, then $w(x, \mu)$ solves problem $(11.2.1_0)$ and $w \not\equiv 0$ in D, because if $w = 0$ in D, then $w_N^+ = 0$, so $w_N^- = 0$ by formula (11.1.32), and, consequently, $w = 0$ in D'. This implies $\mu = w^- - w^+ = 0$, contrary to our assumption.

Assume now that problem $(11.2.1_0)$ has m linearly independent solutions. If $w(x, \mu)$ is known, then (11.2.56) is an equation for σ. The operator Q is symmetric. Therefore,

$$Q^* = \overline{Q}, \tag{11.2.57}$$

where Q^* is the adjoint operator and $\overline{Q}$ is the operator with the complex conjugate kernel. The operator $Q : H^0 \to H^1$ is of Fredholm-type, so equation (11.2.56) is solvable if and only if

$$\int_S (A'\mu - \mu)\overline{\varphi_j} ds = 0, \quad \forall \varphi_j \in N(Q^*). \tag{11.2.58}$$

If $Q^*\varphi_j = 0$, then, by formula (11.2.57), $Q\overline{\varphi_j} = 0$. Thus, condition (11.2.58) can be written as

$$\int_S (A'\mu - \mu)\psi_j ds = 0, \quad \forall \psi_j \in N(Q), \quad \psi_j = \overline{\varphi_j}. \tag{11.2.59}$$

This formula can be rewritten as

$$\int_S w^+ \overline{\varphi_j} ds = 0. \tag{11.2.60}$$

Since $u(x, \overline{\varphi_j})|_S = Q(\overline{\varphi_j}) = 0$, it follows, as before, that $u(x, \overline{\varphi_j}) = 0$ in D', so

$$\overline{\varphi_j} = \frac{\partial u}{\partial N^+}. \tag{11.2.61}$$

Therefore, formula (11.2.60) takes the form

$$\int_S w^+ \frac{\partial u}{\partial N^+} ds = \int_S w_N^+ u \, ds = 0, \tag{11.2.62}$$

because $u|_S = 0$. Therefore, the necessary and sufficient condition for the solvability of equation (11.2.56) when μ is given is satisfied and, consequently, each $w(x, \mu)$ can be represented as $u(x, \sigma)$ in D.

Conversely, if $u(x, \sigma)$ is given in D, then equation (11.2.56) is an equation for μ. This equation is solvable if and only if

$$\int_S Q(\sigma) \overline{h_j} ds = 0, \quad \forall h_j \in N((A' - I)^*) = N(\overline{A} - I). \tag{11.2.63}$$

If $\overline{A} h_j - h_j = 0$, then $A \overline{h_j} - \overline{h_j} = 0$. Consider $u(x, \overline{h_j})$. Then

$$u_N^- = \frac{A \overline{h_j} - \overline{h_j}}{2} = 0. \tag{11.2.64}$$

Therefore, $u(x, \sigma) = 0$ in D', so $Q(\overline{h_j}) = 0$.
Thus, one gets

$$\int_S Q(\sigma) \overline{h_j} ds = \int_S \sigma Q(\overline{h_j}) ds = 0. \tag{11.2.65}$$

Here we have used the symmetry of the kernel of the operator Q. Formula (11.2.65) shows that the necessary and sufficient condition for the solvability of equation (11.2.56) for μ is satisfied. Therefore, given $u(x, \sigma)$ in D, we have found $w(x, \mu)$ such that $u(x, \sigma) = w(x, \mu)$ in D.

Let us prove that for a given $u(x, \sigma)$ in D the $w(x, \mu)$, such that $u = w$ in D, is uniquely defined. Suppose the contrary. Then $u = w_1(x, \mu_1)$, $u = w_2(x, \mu_2)$ in D, so $w_1 = w_2$ in D. This implies that $\mu_1 = \mu_2$. Indeed, $w(x, \mu_1 - \mu_2) = 0$ in D. Thus, $w_N^+ = 0$, so $w_N^- = 0$ by formula (11.1.32). Therefore, $w(x, \mu_1 - \mu_2) = 0$ in D'. Consequently, $\mu_1 - \mu_2 = w^- - w^+ = 0$.
Theorem 11.2.1 is proved. $\square$

Theorem 11.2.2. *If* $u := u(x, \sigma)$ *is given in* D', *then* $w := w(x, \mu)$ *exists such that* $u = w$ *in* D' *if and only if*

$$\int_S u p_j ds = 0, \quad \forall p_j \in N(I + A). \tag{11.2.66}$$

If $w = w(x, \mu)$ *is given in* D', *then* $u = u(x, \mu)$ *exists such that* $u = w$ *in* D' *if and only if*

$$\int_S w^- \varphi_j ds = 0, \quad \forall \varphi_j \in N(Q). \tag{11.2.67}$$

Proof. If $u = w$ in D', then

$$\frac{A'\mu + \mu}{2} = u|_S = Q\sigma. \tag{11.2.68}$$

If u is given in D', then (11.2.68) is an equation for μ. This is a Fredholm-type equation. It is solvable for μ if and only if

$$\int_S u \overline{q_j} ds = 0, \quad \forall q_j \in N(A'^* + I) = N(\overline{A} + I). \tag{11.2.69}$$

If $q_j \in N(\overline{A} + I)$, then $p_j = \overline{q_j} \in N(A + I)$. This and (11.2.69) yield the necessary and sufficient condition (11.2.66) for $u = w$ in D'.

If $w = w(x, \mu)$ is given in D' and $w = u$ in D', then equation (11.2.68) is an equation for σ. This equation is solvable for σ if and only if

$$\int_S \frac{A'\mu + \mu}{2} \overline{r_j} ds = 0, \quad \forall r_j \in N(Q^*). \tag{11.2.70}$$

Equation (11.2.70) can be written as

$$\int_S w^- \overline{r_j} ds = 0, \quad \forall r_j \in N(Q^*). \tag{11.2.71}$$

The kernel of the operator Q is symmetric. Therefore, $Q^* = \overline{Q}$, and $\overline{r_j} \in N(Q)$. Therefore, condition (11.2.71) is identical with the condition (11.2.67).

Theorem 11.2.2 is proved. $\qquad\qquad\qquad\qquad\qquad\qquad\qquad\qquad\qquad\square$

In the proof of Theorem 11.2.1, the following result was used

Lemma 11.2.4. *The operator $Q : H^0 \to H^1$ is of Fredholm type.*

Proof.   One has $Q = Q_0 + Q_1$, where the kernel of the operator Q_0 is $\frac{1}{4\pi r_{st}}$ and the kernel of the operator Q_1 is $\frac{e^{ikr_{st}}-1}{4\pi r_{st}}$. The operator Q_0 is an isomorphism between H^0 and H^1, and the operator Q_1 is compact from H_0 into H_1. Thus,

$$Q = Q_0 + Q_1 \tag{11.2.72}$$

is an operator of Fredholm type.

It is easily checked that Q_1 is compact from H^0 into H^1. Let us check that $Q_0 : H^0 \to H^1$ is an isomorphism. The injectivity of Q_0 follows from the positivity of the Fourier transform:

$$\int_{\mathbb{R}^3} \frac{e^{ix\cdot\xi}}{4\pi|x|}dx = \frac{1}{|\xi|^2}. \tag{11.2.73}$$

This positivity guarantees the positivity of Q_0:

$$(Q_0\sigma, \sigma)_{L^2(S)} > 0, \quad \text{if } \sigma \neq 0. \tag{11.2.74}$$

Indeed, using Parseval's equality, one gets

$$(Q_0\sigma, \sigma)_{L^2(S)} = \frac{1}{(2\pi)^3}\int_{\mathbb{R}^3} \frac{|\widetilde{\sigma_S}(\xi)|^2}{|\xi|^2}d\xi. \tag{11.2.75}$$

From (11.2.75), formula (11.2.74) follows.

The surjectivity of $Q_0 : H^0 \to H^1$ and the boundedness of $Q_0 : H^0 \to H^1$ can be proved by several ways. One of the ways is based on the elliptic estimates for the Laplacian. Namely, let $f \in H^1(S) := H^1$. If $Q^0\sigma = f$, then

$$u(x,\sigma)|_S = f, \quad \nabla^2 u = 0 \quad \text{in } D. \tag{11.2.76}$$

The known estimates for the solutions to (11.2.76) imply the existence of the unique solution, which gives the surjectivity of Q_0 and also its injectivity, and the estimate (Gilbarg and Trudinger (1983), Hörmander (1983–1985)):

$$\|u\|_{H^{3/2}(D)} \leq c\|u\|_{H^1(S)}. \tag{11.2.77}$$

Therefore, the first derivatives of u belong to $H^{1/2}(D)$. By the trace theorem, their restrictions to S belong to $H^0 = L^2(S)$. Since σ can be computed by the formula

$$\sigma = u_N^+ - u_N^-, \tag{11.2.78}$$

it follows that $\sigma \in H^0$. This proves surjectivity of $Q_0 : H^0 \to H^1$ and also its continuity and the continuity of its inverse.

Lemma 11.2.4 is proved. □

11.3 Asymptotic behavior of the solution to the Helmholtz equation under the impedance boundary condition

Consider the following problem:

$$\nabla^2 u + k^2 u = 0 \quad \text{in } D', \tag{11.3.1}$$

$$u_N = \zeta u \quad \text{on } S, \quad \operatorname{Im}\zeta \le 0, \tag{11.3.2}$$

$$u = u_0 + v, \quad v_r - ikv = o\left(\frac{1}{r}\right), \quad r := |x| \to \infty. \tag{11.3.3}$$

Here $u_0 = e^{ik\alpha \cdot x}, \alpha \in S^2, S^2$ is the unit sphere in $\mathbb{R}^3$, $D' = \mathbb{R}^3 \setminus D$, D is a bounded domain with a smooth boundary, N is the unit normal to S pointing out of D. The questions we study in this section are as follows:

Q1: *Does the solution to problems (11.3.1)–(11.3.3) depend continuously on ζ as $|\zeta| \to \infty$?*

In other words, is it true that

$$\lim_{\zeta \to \infty} u = w, \tag{11.3.4}$$

where w solves the problem

$$(\nabla^2 + k^2)w = 0 \quad \text{in } D', \tag{11.3.5}$$

$$w = 0 \quad \text{on } S, \tag{11.3.6}$$

$$w = u_0 + v_1, \tag{11.3.7}$$

where v_1 satisfies the radiation condition (11.3.3).

Q2: *At what rate does the convergence in (11.3.4) hold?*

We prove that (11.3.4) holds and give the rate of the convergence.

The expression $|\zeta| \to \infty$ means throughout that $\mathrm{Re}\zeta \to +\infty$ and $\mathrm{Im}\zeta \leq 0$ is bounded.

Let us look for the unique solution to problems (11.3.1)–(11.3.3) of the form

$$u = u_0 + \int_S g(x,t)\sigma(t)dt, \quad g(x,t) := \frac{e^{ik|x-t|}}{4\pi|x-t|}. \tag{11.3.8}$$

Then equation (11.3.1) and condition (11.3.3) are satisfied with any σ, so that (11.3.8) solves problems (11.3.1)–(11.3.3) if σ is chosen so that the impedance boundary condition (11.3.2) is satisfied.

To satisfy (11.3.2), σ has to satisfy the following integral equation:

$$\frac{A\sigma - \sigma}{2} - \zeta Q\sigma = -u_{0N} + \zeta u_0, \tag{11.3.9}$$

where A is defined in (11.1.5) and

$$Q\sigma := \int_S g(s,t)\sigma(t)dt. \tag{11.3.10}$$

Let us denote

$$\frac{1}{\zeta} := h, \quad h = \frac{\zeta_1 - i\zeta_2}{|\zeta|^2}, \quad \mathrm{Im}h \geq 0, \tag{11.3.11}$$

and rewrite equation (11.3.9) as

$$Q\sigma = h\frac{A\sigma - \sigma}{2} + hu_{0N} - u_0. \tag{11.3.12}$$

Lemma 11.3.1.

$$\sigma = -Q^{-1}u_0 + O(h), \quad h \to 0, \tag{11.3.13}$$

provided that Q is an isomorphism of H^l onto H^{l+1}.

Here $H^l := H^l(S) := W^{l,2}(S)$ is the Sobolev space, $Q : H^l \to H^{l+1}$ is an isomorphism if and only if k^2 is not a Dirichlet eigenvalue of the Laplacian in D, and l can be an arbitrary real number if $S \in C^\infty$.

From (11.3.12), one derives

$$(Q\sigma, \sigma)_l \leq \frac{h}{2}(A\sigma, \sigma)_l + h(u_{0N}, \sigma)_l - (u_0, \sigma)_l, \quad l \geq 0. \quad (11.3.14)$$

The idea of the further argument is to derive (11.3.13) from (11.3.14).

Proof of Lemma 11.3.1. The operator $A : H^l \to H^{l+1}$ is continuous. The form $(Q\sigma, \sigma)$ can be estimated from above and below

$$c_1||\sigma||_{l-\frac{1}{2}} \leq (Q\sigma, \sigma)_l \leq c_2||\sigma||_{l-\frac{1}{2}}, \quad (11.3.15)$$

because $Q : H^l \to H^{l+1}$ is an isomorphism.

Thus, estimate (11.3.14) implies

$$||\sigma||^2_{l-\frac{1}{2}} \leq ch||\sigma||^2_{l-\frac{1}{2}} + ch||\sigma||_{l-\frac{1}{2}}||u_{0N}||_{l+\frac{1}{2}} + ||u_0||_{l+\frac{1}{2}}||\sigma||_{l-\frac{1}{2}}. \quad (11.3.16)$$

For small h, this estimate implies

$$||\sigma||_{l-\frac{1}{2}} \leq ||u_0||_{l+\frac{1}{2}} \leq c(l), \quad (11.3.17)$$

since $u_0 \in H^l$ for any l.

Take $l = 1/2$. Then estimate (11.3.17) shows that

$$||\sigma||_0 \leq c_0, \quad ||\sigma||_1 \leq c_1, \quad (11.3.18)$$

where c_0 and c_1 do not depend on h. When $h \to 0$, one can select a sequence $\sigma(h)$ convergent in H^0 a to an element σ_0. The corresponding element $\sigma_0 = -Q^{-1}u_0$, as follows from equation (11.3.12). The rate of convergence is $O(h)$ as $h \to 0$.

Lemma 11.3.1 is proved. $\square$

Lemma 11.3.2. *Estimate*

$$c_1||\sigma||_{l-\frac{1}{2}} \leq |(Q\sigma, \sigma)| \leq c_2||\sigma||_{l-\frac{1}{2}}, \quad c_1, c_2 = \text{const} > 0, \quad (11.3.19)$$

holds.

Proof. It is sufficient to check estimate (11.3.19) for the operator

$$Q_0\sigma := \int_S \frac{\sigma(t)dt}{4\pi r_{st}}, \quad (11.3.20)$$

because $Q = Q_0 + Q_1$, Q_1 is smooth, $Q_1 : H^l \to H^{l+1}$, and Q is assumed to be an isomorphism of H^l onto H^{l+1}. That Q_1 is smooth is clear since its kernel is

$$\frac{e^{ikr_{st}}}{4\pi r_{st}} = \frac{ik}{4\pi} + O(r_{st}). \tag{11.3.21}$$

The operator $Q_0 : H^l \to H^{l+1}$ is an isomorphism of H^l onto H^{l+1}, as we have proved in the previous section. It is a positive self-adjoint operator in H^0, and

$$c_1 \|\sigma\|_{l-\frac{1}{2}}^2 \leq (Q_0\sigma, \sigma)_l \leq c_2 \|\sigma\|_{l-\frac{1}{2}}^2. \tag{11.3.22}$$

Therefore, one has

$$Q = Q_0(I + Q_0^{-1}Q_1). \tag{11.3.23}$$

The operator $Q_0^{-1}Q_1$ is compact in H^l because $Q_0^{-1}Q_1 : H^l \to H^{l+1}$, and the embedding operator $i : H^{l+1} \to H^l$ is compact.

Since Q and Q_0 are isomorphisms from H^l onto H^{l+1}, it follows that $I + Q_0^{-1}Q_1$ is an isomorphism of H^l onto H^l.

Therefore, estimate (11.3.19) holds.

Lemma 11.3.2 is proved.   □

Lemma 11.3.3. *The solution to equation (11.3.12) depends continuously on the parameter h in the space $H^l, l \geq 0$, provided that $S \in C^\infty$.*

Proof. We have proved the above statement for $h = 0$. For any other h, $\mathrm{Im}\, h \geq 0$, the proof is similar.

Lemma 11.3.3 is proved.   □

Let us formulate the result of this section as a theorem.

Theorem 11.3.1. *Assume that $S \in C^\infty$ and $\mathrm{Im}\, \zeta \leq 0$. Then problems (11.3.1)–(11.3.3) have a solution, this solution is unique and it depends continuously on ζ. In particular,*

$$u = w + O(|h|), \quad |h| \to 0. \tag{11.3.24}$$

Here w solves the problems (11.3.5)–(11.3.7) and $h = \frac{1}{\zeta}$.

11.4 Some properties of the electrical capacitance

If $D \subset \mathbb{R}^3$ is a bounded domain with the boundary S, then the problem

$$\Delta u = 0 \quad \text{in } D' := \mathbb{R}^3 \backslash D, \tag{11.4.1}$$

$$u|_S = 1, \tag{11.4.2}$$

$$u(\infty) = 0, \tag{11.4.3}$$

is the problem perfect conductor D, charged to a potential $U = 1$. The surface charge density $\sigma = \sigma(t)$ generates the total charge Q as follows:

$$Q = \int_S \sigma(t)dt, \tag{11.4.4}$$

dt is the element of the surface area. The total charge Q is proportional to the potential U to which the perfect conductor is charged:

$$Q = CU. \tag{11.4.5}$$

If $U = 1$, then $Q = C$. The coefficient C is called electrical capacitance of the conductor D. The energy of this conductor is

$$E = \frac{CU^2}{2} = \frac{QU}{2}. \tag{11.4.6}$$

This formula, known from the middle school physics course, can be derived by using the expression

$$E = \frac{1}{2} \int_{D'} E^2(x)dx = \frac{1}{2} \int_{D'} |\nabla u|^2 dx$$

$$= -\frac{1}{2} \int_{D'} u \Delta u \, dx - \frac{1}{2} \int_S u u_N dx = \frac{Q}{2}, \tag{11.4.7}$$

where we took into account the equations

$$\Delta u = 0 \quad \text{in } D', \quad u|_S = 1, \quad -u_N(t)|_S = \sigma(t), \quad \int_S \sigma(t)dt = Q, \tag{11.4.8}$$

and N is the unit normal to S pointing out of D.

Formula (11.4.7) coincides with (11.4.6) because $U = 1$. Since the energy attains its minimum among energies corresponding to all the functions u in D', $u \in H^1(D')$, such that $u|_S = 1$ and $u(\infty) = 0$, one may write

$$C = \min_{u|_S=1, u(\infty)=0} \int_{D'} |\nabla u|^2 dx. \tag{11.4.9}$$

By $H^1(D')$, the space of functions for which the integral (11.4.9) is finite is denoted. The surface charge distribution $\sigma(t)$, corresponding to the minimizer of the integral (11.4.9), satisfies the equation

$$u(s) = \int_S \frac{\sigma(t)dt}{4\pi|s-t|} = 1, \tag{11.4.10}$$

which shows that the conductor D is charged to the potential $U = 1$.

The Euler–Lagrange equation for the minimizer of (11.4.9) is the equation

$$\Delta u = 0 \quad \text{in } D'. \tag{11.4.11}$$

From the definition (11.4.9), the following result can be derived.

Lemma 11.4.1. *If $D_1 \subseteq D_2$, then $C_1 \leq C_2$.*

Proof. If $D_1 \subset D_2$, then $D_1' \supset D_2'$. Therefore, minimization in (11.4.9) is conducted over a larger set if $D_1 \subseteq D_2$. Consequently, $C_1 \leq C_2$. Lemma 11.4.1 is proved. $\qquad\square$

The reader can prove the strict inequality: $C_1 < C_2$ if $D_1 \subset D_2$.

Let us prove the following variational principle for C:

$$C = \max_{\sigma \in L^2(S)} \frac{\left|\int_S \sigma(t)dt\right|^2}{\int_S \int_S \frac{\sigma(t)\sigma(s)dsdt}{4\pi|s-t|}}. \tag{11.4.12}$$

This variational principle can be written as

$$C^{-1} = \min_{\sigma \in L^2(S)} \frac{\int_S \int_S \frac{\sigma(t)\sigma(s)dsdt}{4\pi|s-t|}}{\left|\int_S \sigma(t)dt\right|^2}. \tag{11.4.13}$$

In the form (11.4.13), the variational principle yields Gauss's principle:

The minimal value of the functional in (11.4.13) is C^{-1} and is attained at the solution σ of equation (11.4.10).

If one takes $\sigma = 1$ in (11.4.12), then one gets

$$C \geq \frac{4\pi|S|^2}{J}, \quad J := \int_S \int_S \frac{dsdt}{|s-t|}, \tag{11.4.14}$$

and $|S|$ denotes the surface area of S. We derive principle (11.4.12) as a particular case of a more general principle.

Suppose that $A = A^*$ is a self-adjoint operator in a Hilbert spact H. Consider the equation

$$Au = f. \tag{11.4.15}$$

Theorem 11.4.1. *Formula*

$$(Au, u) = \max_{v \in H} \frac{|(Au, v)|^2}{(Av, v)} \tag{11.4.16}$$

holds if and only if $A \geq 0$.

Recall that $A \geq 0$ if and only if $(Av, v) \geq 0$ for every $v \in H$. If $(Av, v) = 0$ in (11.4.16), then we define the ratio in (11.4.16) to be equal to zero. Let us also recall that if $A \geq 0$, then the Cauchy inequality holds as follows:

$$|(Au, v)|^2 \leq (Au, u)(Av, v), \tag{11.4.17}$$

and the equality sign in (11.4.17) is attained if and only if u and v are linearly dependent.

Proof of Theorem 11.4.1. The sufficiency of the condition $A \geq 0$ follows from the inequality (11.4.17). Indeed, (11.4.17) implies

$$\frac{|(Au, v)|^2}{(Av, v)} \leq (Au, u), \tag{11.4.18}$$

and the equality sign in (11.4.18) is attained if $v = \lambda u$, $\lambda =$ const.

Let us prove the necessity of the condition $A \geq 0$. This requires more work. First, note that (Av, v) cannot be negative for all $v \in H$, because in this case (11.4.16) implies a contradiction: multiply (11.4.16) by -1 and take into account that $-A := B \geq 0$ if $(Av, v) \geq 0$ for every $v \in H$. Then (11.4.16) yields

$$(-Au, u) = -\max_{v} \frac{|(-Au, v)|^2}{(Av, v)} = \min_{v} \frac{|(-Au, v)|^2}{(-Av, v)}. \qquad (11.4.19)$$

Consequently, with $-A = B \geq 0$, one has

$$(Bu, u)(Bv, v) \leq |(Bv, u)|^2, \quad \forall u, v \in H. \qquad (11.4.20)$$

This contradicts the Cauchy inequality (11.4.17) because $B \geq 0$. Thus, $(Aw, w) > 0$ for some w and formula (11.4.16) implies that $(Au, u) \geq 0$ for all u.

Alternatively, we can assume that (11.4.16) is valid, but there are two elements z and w such that $(Az, z) > 0$ and $(Aw, w) < 0$, and show that this is impossible.

Take $v = \lambda z + w$, where λ is an arbitrary real number. Then (11.4.16) yields

$$\frac{|(Au, \lambda z + w)|^2}{q(\lambda)} \leq (Au, u), \qquad (11.4.21)$$

where

$$q(\lambda) = a\lambda^2 + 2b\lambda + c, \quad a := (Az, z) > 0, \quad c = (Aw, w) < 0, \qquad (11.4.22)$$

and $b = \mathrm{Re}(Az, w)$. The polynomial $q(\lambda)$ has, therefore, two roots, $\lambda_1 < 0$ and $\lambda_2 > 0$, and $q^{-1}(\lambda) \to +\infty$ if $\lambda \to \lambda_1 - 0$ or $\lambda \to \lambda_2 + 0$. The quadratic polynomial $p(\lambda) := |(Au, \lambda z + w)|^2$ has also two roots which have to coincide with λ_1 and λ_2 because the ratio on the left of (11.4.21) is bounded from above. Since $\lambda_1, \lambda_2 < 0$ and λ is a real number, one calculates $p(\lambda)$ by the following formula:

$$p(\lambda) = \lambda^2 |(Au, z)|^2 + 2\lambda \mathrm{Re}(Au, z)\overline{(Au, w)} + |(Au, z)|^2.$$

If this polynomial has two roots of different signs, then

$$\frac{|(Au, w)|^2}{|(Au, z)|^2} < 0. \tag{11.4.23}$$

Inequality (11.4.23) is a contradiction.

Theorem 11.4.1 is proved. □

To derive from Theorem 11.4.1 principle (11.4.12), one uses equation (11.4.10), so that

$$A\sigma = \int_S \frac{\sigma(t)dt}{4\pi|s - t|} = 1. \tag{11.4.24}$$

We will check in the following that $A \geq 0$. Principle (11.4.16) yields

$$\int_S \sigma(t)dt = \max_{v \in L^2(S)} \frac{\left|\int_S v(t)dt\right|^2}{\int_S \int_S \frac{v(t)v(s)dsdt}{4\pi|s-t|}}, \tag{11.4.25}$$

where σ is the unique solution to equation (11.4.24), and, therefore, $\int_S \sigma(t)dt = Q = C$, so that principle (11.4.12) is proved, provided that one checks that the operator A in (11.4.24) satisfies condition $A \geq 0$.

Let us check this.

Lemma 11.4.2. *The operator A in (11.4.24) satisfies the condition $A \geq 0$.*

Proof of Lemma 11.4.2. Applying Parseval's equality to the quadratic form

$$(Av, v) = \int_S \int_S \frac{v(t)v(s)dsdt}{4\pi|s - t|} \tag{11.4.26}$$

and using the Fourier transform

$$F\left(\frac{1}{|x|}\right) = \int_{\mathbb{R}^3} \frac{e^{-i\xi \cdot x}}{4\pi|x|}dx = \frac{1}{|\xi|^2} > 0, \tag{11.4.27}$$

one gets

$$(Av, v) = \frac{1}{(2\pi)^3} \int_{\mathbb{R}^3} \frac{|\tilde{\sigma}_S(\xi)|^2}{|\xi|^2}d\xi > 0, \tag{11.4.28}$$

where

$$\tilde{\sigma}_S(\xi) = \int_{\mathbb{R}^3} e^{-i\xi \cdot x} \sigma(s) \delta_S(x) dx = \int_S e^{-i\xi \cdot s} \sigma(s) ds. \qquad (11.4.29)$$

Here $\delta_S(x)$ is the delta-function concentrated on the surface S.
Lemma 11.4.2 is proved. $\qquad\qquad\qquad\qquad\qquad\qquad\qquad\qquad\square$

11.5 Summary of the results

The basic results in this chapter are formulated in Theorems 11.1.1, 11.2.1, 11.2.2, 11.3.1, 11.4.1.

Chapter 12

Collocation Method

12.1 Convergence of the collocation method

In this book on several occasions the following statement has been used:

The linear algebraic system (LAS):

$$u_q = u_{0q} - \sum_{p \neq q}^{P} g(x_q, x_p) h(x_p) N(x_p) u(x_p) |\Delta_p| \qquad (12.1.1)$$

in the limit $a \to 0$ becomes the integral equation

$$u(x) = u_0(x) - \int_\Omega g(x, y) h(y) N(y) u(y) dy, \qquad (12.1.2)$$

where Δ_p is a cube, $|\Delta_p|$ is its volume, which tends to zero as $a \to 0$; the union of the cubes $|\Delta_p|$ contains Ω and its volume tends to the volume of Ω as $a \to 0$; the cubes $|\Delta_p|$ and $|\Delta_q|$ have no common interior points if $p \neq q$.

More precisely:

If equation (12.1.2) is uniquely solvable for any $u_0 \in C(\Omega)$, then LAS (12.1.1) is uniquely solvable for all sufficiently small $a > 0$ and the limit in $C(\Omega)$ of the function

$$u_M(x) := \sum_{p=1}^{M} u(x_p) \chi_{\Delta_p}(x) \qquad (12.1.3)$$

equals to the unique solution of equation (12.1.2). By χ_{Δ_p}, the characteristic function of Δ_p is denoted.

The aim of this chapter is to justify the above statements.

Consider an integral equation

$$Au := u + Tu = f, \tag{12.1.4}$$

where

$$Tu = \int_\Omega T(x,y)u(y)dy, \tag{12.1.5}$$

$\Omega \subset \mathbb{R}^3$ is a bounded domain and T is a linear compact operator in $X = L^\infty(\Omega)$. Let us assume that

$$N(A) := \{u : Au = 0\} = \{0\}. \tag{12.1.6}$$

Since $A = I + T$ is a Fredholm-type operator, assumption (12.1.6) implies that A is a bijection of X onto X and $\|A\| \le c$. Equation (12.1.2) is of this type. Condition (12.1.6) is verified as follows. If v is a solution to the homogeneous equation (12.1.2), then, applying the operator $\nabla^2 + k^2$ to (12.1.2) and denoting

$$q(x) := N(x)h(x),$$

one gets

$$\left(\nabla^2 + k^2 - q(x)\right) v = 0 \quad \text{in } \mathbb{R}^3, \tag{12.1.7}$$

and v satisfies the radiation condition

$$v_r - ikv = o\left(\frac{1}{r}\right), \quad r = |x| \to \infty. \tag{12.1.8}$$

Now, by Lemma 2.1.2, one concludes that $v = 0$ provided that $\operatorname{Im} q \le 0$ and q is compactly supported. These conditions are satisfied because $\operatorname{Im} h \le 0$ and $N(x) \ge 0$. Thus, $v = 0$.

Let us assume that

$$\sup_{x \in \Omega} \int_\Omega \left(|T(x,y)| + |\nabla_x T(x,y)|\right) dy := c_T < \infty. \tag{12.1.9}$$

Consider a partition of Ω into a union of small cubes Δ_j with a side $\frac{1}{n}$ and assume that the partition cubes with different indices j have no common interior points. Let j_n denote the total number of the partition cubes. Then

$$\Omega \subset \bigcup_{j=1}^{j_n} \Delta_j, \quad \text{diam} \Delta_j = b_n = \frac{\sqrt{3}}{n}, \quad j_n = O\left(n^3\right). \qquad (12.1.10)$$

Let x_j be the center of the cube Δ_j, and $\chi_j(x)$ be the characteristic function of Δ_j:

$$\chi_j(x) = \begin{cases} 1 & \text{in } \Delta_j, \\ 0 & \text{in } \Delta_j' = \mathbb{R}^3 \backslash \Delta_j. \end{cases} \qquad (12.1.11)$$

By $\omega_u(\delta)$, we denote the *modulus of continuity* of a function $u(x)$, as follows:

$$\omega_u(\delta) := \sup_{|x-y| \leq \delta, x,y \in \Omega} |u(x) - u(y)|. \qquad (12.1.12)$$

By $C^a(\Omega)$, $a \in (0,1]$, let us denote the set of functions such that

$$\omega_u(\delta) \leq c\delta^a. \qquad (12.1.13)$$

Consider the following LAS:

$$u_i + \sum_{j=1}^{j_n} T_{ij} u_j = f_i, \quad 1 \leq i \leq j_n, \qquad (12.1.14)$$

where

$$u_i := u(x_i), \quad T_{ij} := \int_{\Delta_j} T(x_i, y) dy; \quad f_i := f(x_i). \qquad (12.1.15)$$

Equation (12.1.14) is in one-to-one correspondence with the following equation in X:

$$u_i(x) + \sum_{j=1}^{j_n} \chi_i(x) \int_{\Delta_j} T(x_i, y) u(y) dy = f_i(x), \quad 1 \leq i \leq j_n, \qquad (12.1.16)$$

where

$$u_i(x) = u_i \chi_i(x). \tag{12.1.17}$$

Let

$$u^{(n)}(x) := \sum_{i=1}^{j_n} u_i(x) := \sum_{i=1}^{j_n} u_i \chi_i(x). \tag{12.1.18}$$

Equation (12.1.16) is equivalent to the following equation:

$$(I + T_n) u^{(n)} = f^{(n)} \tag{12.1.19}$$

where $f^{(n)}$ is defined as in (12.1.18) and

$$T_n u^{(n)} := \int_\Omega T^{(n)}(x, y) u^{(n)}(y) dy, \quad T^{(n)}(x, y) := \sum_{i=1}^{n} \chi_i(x) T(x_i, y). \tag{12.1.20}$$

Let us formulate our first result.

Lemma 12.1.1. *Equation* (12.1.19) *in* X *is equivalent to equation* (12.1.14) *in* $\mathbb{R}^{j_n}$ *in the following sense:*

If $\{u_i\}_{i=1}^{j_n}$ *solves* (12.1.14), *then the function* (12.1.18) *solves* (12.1.19), *and conversely, if* (12.1.18) *solves* (12.1.19), *then* $\{u_i\}_{i=1}^{j_n}$ *solves* (12.1.14).

Proof. Assume that $\{u_i\}_{i=1}^{j_n}$ solves (12.1.14). Multiply (12.1.14) by $\chi_i(x)$ and sum up with respect to i from 1 to j_n. Using definition (12.1.18) we obtain equation (12.1.19) with T_n defined in (12.1.20).

Conversely, assume that equation (12.1.19) has a solution (12.1.18). Set $x = x_i$ in equation (12.1.19), and use the relation

$$\chi_j(x_i) = \delta_{ij} = \begin{cases} 0 & \text{if } i \neq j, \\ 1 & \text{if } i = j, \end{cases} \tag{12.1.21}$$

to get equation (12.1.14) with T_{ij} defined in (12.1.15).

Lemma 12.1.1 is proved. $\square$

Theorem 12.1.1. *If assumption* (12.1.9) *holds, then LAS* (12.1.14) *has a unique solution for all n sufficiently large and this solution generates by formula* (12.1.18) *the function* $u^{(n)}$ *such that*

$$\left| u^{(n)}(x) - u(x) \right| \le c\omega_u \left(\frac{1}{n} \right) \to 0 \quad as \ n \to \infty, \qquad (12.1.22)$$

where u solves equation (12.1.4).

Solving operator equation (12.1.4) by reducing it to solving LAS (12.1.14) is a version of the *collocation method.* Our results and their proofs remain valid for solving the equation

$$Bu + Tu = f, \qquad (12.1.23)$$

where B is an isomorphism of S, T is a linear compact operator, and

$$N(B + T) = \{0\}.$$

The usual collocation method for solving equation (12.1.4) consists of the following. Let $X_n \subset X_{n+1}$, $\dim X_n = n$, be a limit-dense sequence of subspaces in X, that is, $\lim_{n \to \infty} \rho(u, X_n) = 0$, where $\rho(u, X_n)$ is the distance from the element u to the subspace X_n. Let $\{w_i\}_{i=1}^n$ be a basis of X_n. Choose $x_j \in \Omega$, $1 \le j \le n$, such that

$$\det w_i(x_j) \neq 0, \quad 1 \le i, j \le n. \qquad (12.1.24)$$

Then the linear algebraic system

$$\sum_{i=1}^{n} w_i(x_j) c_i = f_j, \quad 1 \le j \le n, \qquad (12.1.25)$$

is uniquely solvable for c_i for any f_j.

The collocation method for solving equation (12.1.4) consists of looking for an approximate solution to (12.1.4) of the form

$$u_n = \sum_{i=1}^{n} c_i^{(n)} w_i(x), \quad c_i^{(n)} = \text{const}, \qquad (12.1.26)$$

and finding coefficients $c_i^{(n)}$ from the system

$$\sum_{i=1}^{n} c_i^{(n)} w_i(x_j) + \sum_{i=1}^{n} c_i^{(n)} \int_\Omega T(x_j, y) w_i(y) dy = f(x_j), \quad 1 \le j \le n.$$

$$(12.1.27)$$

In this version of the collocation method, the points x_j are chosen so that the system (12.1.27) is uniquely solvable for $c_i^{(n)}$. The basis functions $w_i(x)$ are often taken to be algebraic or trigonometric polynomials. The projection operator on the subspace of polynomials of degree $\le n$ has a norm in $C(\Omega)$ that tends to infinity as $n \to \infty$, see Kantorovich and Akilov (1982).

In our version of the collocation method, the choice of $x_j \in \Delta_j$ is arbitrary, the basis functions are $\chi_i(x)$, and the norm of the projection operators P_n on the space X_n which has basis $\chi_i(x)$, $1 \le i \le n$, does not grow to infinity as $n \to \infty$, it stays ≤ 1. The kernel $T(x, y)$ satisfies our assumption (12.1.9) even if it is unbounded at $x = y$.

For example, if $T(x, y) = \frac{e^{ik|x-y|}}{4\pi|x-y|} q(y)$, where $q \in L^\infty(\Omega)$, as in our example (12.1.2). Indeed, $\nabla_x g(x, y) = O\left(\frac{1}{|x-y|^2}\right)$, and $\sup_{x \in \Omega} \int_\Omega \frac{dy}{|x-y|^2} < \infty$.

One can apply general theorems about convergence of the projection methods to a study of convergence of the collocation method (see Kantorovich and Akilov (1982), Ramm (1986), etc).

In equation (12.1.2), the function $u_0 = e^{ik\alpha \cdot x}$ is smooth and $\omega_{u_0}\left(\frac{1}{n}\right) \le \frac{c}{n}$.

After these remarks let us prove Theorem 12.1.1.

Proof of Theorem 12.1.1. Let us check that the condition

$$\lim_{n \to \infty} ||T_n - T|| = 0, \qquad (12.1.28)$$

where T_n is the operator in equation (12.1.19), allows one to prove estimate (12.1.22). Since we have assumed that $N(I + T) = \{0\}$ and T is compact, the Fredholm alternative guarantees that the operator $(I + T)^{-1}$ is bounded in the operator norm. Since the set of boundedly invertible linear bounded operators is open (see Appendix A), it follows from (12.1.28) that $(I + T_n)^{-1}$ is bounded for all sufficiently large n, $n > n_0$. In particular, $N(I + T_n) = \{0\}$ for $n > n_0$.

By Lemma 12.1.1, equation (12.1.19) is equivalent to LAS (12.1.14). Therefore, this LAS is uniquely solvable for $n > n_0$ and its solution generates the solution $u^{(n)}$ to equation (12.1.19) by formula (12.1.18).

Theorem 12.1.1 will be proved if formula (12.1.28) is established. Indeed,

$$\|u^{(n)} - u\| = \|(I + T_n)^{-1} f^{(n)} - (I + T)^{-1} f\|$$
$$\leq \|(I + T_n)^{-1}(f - f^{(n)})\|$$
$$+ \|(I + T_n)^{-1} - (I + T)^{-1}\| \, \|f^{(n)}\|. \quad (12.1.29)$$

Since

$$\|f - f^{(n)}\| \leq c \omega_f \left(\frac{1}{n}\right) \leq c\frac{1}{n}, \quad n \to \infty, \quad (12.1.30)$$

one has $\sup_n \|f^{(n)}\| < \infty$, and the right-hand side of (12.1.29) tends to zero if

$$\|(I + T_n)^{-1} - (I + T)^{-1}\| = \|(I + T_n)^{-1}(T_n - T)(I + T)^{-1}\|$$
$$\leq c\|T_n - T\| \to 0, \quad (12.1.31)$$

that is, if (12.1.28) holds. Estimate (12.1.30) follows from the definition of $f^{(n)}$, see (12.1.18), and from the assumption that $f \in C^1(\Omega)$. The assumption (12.1.9) yields

$$\int_\Omega \left| T(x, y) - T^{(n)}(x, y) \right| dy \leq \sup_{1 \leq i \leq n} \sup_{x \in \Delta + i} \int_\Omega |T(x, y) - T(x_i, y)| dy$$
$$\leq \frac{c}{n} \int_\Omega |\nabla_x T(x, y)| dy \leq \frac{c c_T}{n}. \quad (12.1.32)$$

This estimate implies (12.1.28). Consequently,

$$\lim_{n \to \infty} \|u^{(n)} - u\| = 0. \quad (12.1.33)$$

Convergence of our collocation method is established. Estimate (12.1.32) implies also estimate (12.1.22). Indeed

$$|u^{(n)}(x) - u(x)| = \left| \sum_{i=1}^{j_n} \chi_i(x) \int_\Omega T(x_i, y)u(y)dy - \int_\Omega T(x, y)u(y)dy \right|$$

$$= \left| \sum_{i=1}^{j_n} \chi_i(x) \int_\Omega [T(x_i, y) - T(x, y)]u(y)dy \right|$$

$$\leq \sup_{1 \leq i \leq j_n} \sup_{x \in \Delta_i} |x_i - x| \sup_{x \in \Omega} \int_\Omega |\nabla_x T(x, y)|dy \sup_{y \in \Omega} |u|$$

$$\leq \frac{cc_T}{n} \sup_{y \in \Omega} |u|. \tag{12.1.34}$$

Thus,

$$\sup_{x \in \Omega} |u^{(n)}(x) - u(x)| \leq \frac{\sqrt{3}c_T}{n}. \tag{12.1.35}$$

Theorem 12.1.1 is proved. □

12.2　Collocation method and homogenization

Our results open a new approach to the homogenization theory. In many works on this theory (for example, in Bensoussan *et al.* (2011), Zhikov *et al.* (1994)) it is assumed that one deals with a partial differential equation whose coefficients depend on a parameter and are periodic, and one is looking for a limiting behavior of the solution as the parameter tends to zero.

The operator in a homogenization problem is considered on a finite cell, that is, on a bounded region, with periodic boundary conditions, so its spectrum is discrete and the considerations use variational type of inequalities, and often use the self-adjointness of the operators involved.

In the version of the theory developed in this book, the periodicity assumption, the discreteness of the spectrum of the operators involved, and their self-adjointness are not assumed and not used.

Our scheme is based on the convergence of the version of the collocation method established in Theorem 12.1.1. The general scheme for the applications of this version of the homogenization method used in this book can be described as follows.

A physical quantity u is related to itself (through a scattering theory or some other physical theory) by an equation

$$u_j = f_j - \sum_{m \neq j} A_{jm} q_m u_m |\Delta_m|, \qquad (12.2.1)$$

where $u_j := u(x_j)$, $x_j \in \Delta_j$, $\Delta_j \subset \Omega$ is a small subdomain of Ω, $\bigcup_{j=1}^{M} \Delta_j \supset \Omega$, Δ_j and Δ_m for $j \neq m$ do not have common interior points, $|\Delta_m|$ is the volume of Δ_m, $A_{jm} := A(x_j, x_m)$, $q_m = q(x_m)$, $f_j = f(x_j)$, where f is a given function. Suppose that the equation

$$u(x) = f(x) - \int_{\Omega} A(x, y) q(y) u(y) dy \qquad (12.2.2)$$

has a unique solution in $X = L^{\infty}(\Omega)$. Then, under the assumptions stated in Theorem 12.1.1, system (12.2.1) is uniquely solvable in X and its solution tends to the solution of the limiting equation (12.2.2), as $M \to \infty$, in the sense specified in Theorem 12.1.1. The solution $u(x)$ of equation (12.2.2) is considered a "homogenized" version of the discrete physical quantity $\{u_j\}_{j=1}^{M}$.

This scheme is fairly general since convergence of the collocation method can be established for equations more general than (12.2.2).

12.3 Summary of the results

In this chapter, the main result is Theorem 12.1.1. It gives sufficient conditions for convergence of the collocation method proposed for solving integral equation (12.1.2) and more general equations. Theorem 12.1.1 is used as a theoretical basis for a justification of the method.

Chapter 13

Some Inverse Problems Related to Small Scatterers

13.1 Finding the position and size of a small body from the scattering data

Consider a system of two bodies one of which, D_1, is large compared with the wavelength and the other one, D_2, which is small. Let a denote the characteristic size of the small body.

Then $ka \ll 1$, where k is the wave number. In applications, one may think that D_1 is a ship and D_2 is a mine.

Let us assume that the body D_1 is known in the sense that the scattering of waves on the body D_1 in the absence of body D_2 is known. The problem we study consists of the following. Suppose we can measure the field scattered by both bodies. The question we raise is the following:

Can one determine the position and size of D_2 given the position of D_1?

Let us formulate this problem precisely. Let

$$\left(\nabla^2 + k^2\right) u = 0 \quad \text{in } \mathbb{R}^3 \backslash D, \quad D := D_1 \cup D_2, \tag{13.1.1}$$

$$u|_S = 0, \quad S = S_1 \cup S_2, \quad S_j = \partial D_j, \quad j = 1, 2, \tag{13.1.2}$$

$$u = u_0 + v, \quad u_0 := e^{ik\alpha \cdot x}, \quad \alpha \in S^2, \tag{13.1.3}$$

where S^2 is the unit sphere in $\mathbb{R}^3$, and v is the scattered field which satisfies the radiation condition

$$v = A(\beta, \alpha, k)\frac{e^{ikr}}{r} + o\left(\frac{1}{r}\right), \quad r := |x| \to \infty, \quad \beta = \frac{x}{r}. \quad (13.1.4)$$

The coefficient $A(\beta, \alpha, k) := A(\beta, \alpha)$ is called the scattering amplitude, $k > 0$ is fixed.

Problems (13.1.1)–(13.1.4) have a solution and this solution is unique. This is proved as in Theorem 1.1.1.

Let d denote the distance between D_1 and D_2. We assume that

$$ka \ll 1, \quad kd \gg 1. \quad (13.1.5)$$

The first assumption means that D_2 is much smaller than the wavelength $\lambda = \frac{2\pi}{k}$, while the second assumption means that the distance d between D_1 and D_2 is much greater than the wavelength.

We assume that the scattering problems (13.1.1)–(13.1.4) for one body D_1, that is, in the case $D = D_1$, is solved and the corresponding Green's function g_1 is known as follows:

$$\left(\nabla^2 + k^2\right) g_1(x, y) = -\delta(x - y) \quad \text{in } D_1' := \mathbb{R}^3 \backslash D, \quad (13.1.6)$$

$$g_1|_{S_1} = 0, \quad (13.1.7)$$

$$\frac{\partial g_1}{\partial |x|} - ikg_1 = o\left(\frac{1}{|x|}\right), \quad |x| \to \infty. \quad (13.1.8)$$

Our *goal* is to give analytical formulas, asymptotically exact as $a \to 0$, which allow one to locate the position of the small body and estimate its size, given the field scattered by the system of two bodies in which the location of D_1 is known.

Let us look for the (unique) solution to problems (13.1.1)–(13.1.4) of the form

$$u(x) = u_1(x) + \int_{S_2} g_1(x, t)\sigma(t)dt, \quad (13.1.9)$$

where $\sigma(t)$ is unknown and $u_1(x)$ is the solution of the scattering problem for D_1 in the absence of D_2. The function u_1 is known by our assumption. Function (13.1.9) solves equation (13.1.1), satisfies condition (13.1.3), and it will be the solution to problems (13.1.1)–(13.1.4) if σ can be chosen so that condition (13.1.2) is satisfied.

These arguments did not use the smallness of D_2. If D_2 is small, then the function (13.1.9) can be written at the far distances from D_1 as

$$u(x) = u_1(x) + g_1(x, \xi)Q + O\left(\frac{a}{d}\right),\qquad(13.1.10)$$

where

$$Q := \int_{S_2} \sigma(t)dt, \quad \xi \in D_2,\qquad(13.1.11)$$

and ξ is an arbitrary point inside D_2. To get the error term $O\left(\frac{a}{d}\right)$ in (13.1.10), one estimates the following integral:

$$\left|\int_{S_2} [g(x, t) - g(x, \xi)]\sigma(t)dt\right| \leq O\left(|\nabla_t g(x, t)| \, aQ\right)$$

$$\leq O\left(\frac{a}{d}\right)Q = O\left(\frac{a}{d}\right),\qquad(13.1.12)$$

and, since Q is bounded when a is bounded and $|\nabla_t g(x, t)| = O\left(\frac{1}{d}\right)$ when $|x - t|$ is large, one gets formula (13.1.10).

Let us calculate the field scattered by two bodies. Take x far from both bodies in the direction β. Then

$$u_1(x) = e^{ik\alpha \cdot x} + A_1(\beta, \alpha)\frac{e^{ik|x|}}{|x|} + o\left(\frac{1}{|x|}\right),\qquad(13.1.13)$$

where $A_1(\beta, \alpha)$ is the scattering amplitude, and

$$g_1(x, \xi) = \frac{e^{ik|x|}}{|x|}\phi(\xi, -\beta, k) + o\left(\frac{1}{|x|}\right), \quad |x| \to \infty, \quad \frac{x}{|x|} = \beta.$$
$$(13.1.14)$$

This formula is proved by the author in Ramm (1986), p. 46. The function $\phi(\xi, -\beta, k)$ is the solution to the problem

$$\left(\nabla^2 + k^2\right)\phi = 0 \quad \text{in } D_1' := \mathbb{R}^3 \backslash D_1,\qquad(13.1.15)$$

$$\phi|_{S_1} = 0,\qquad(13.1.16)$$

$$\phi = e^{-ik\beta \cdot x} + \psi,\qquad(13.1.17)$$

where ψ satisfies the radiation condition. Note that ϕ is constructed as the scattering solution corresponding to the incident direction $-\beta$ when the limit in (13.1.14) is taken along the direction β.

From (13.1.10) and (13.1.14) it follows that the field, scattered by the system of two bodies, is

$$\frac{e^{ik|x|}}{|x|}\left(\phi(\xi, -\beta, k)Q + A_1(\beta, \alpha)\right). \tag{13.1.18}$$

The function $\phi(\xi, -\beta, k)$ is known if ξ is known because ϕ is defined by the known function g_1. Let us find an analytical formula for Q. The exact boundary condition on S_2 is (13.1.2). This and (13.1.9) yield

$$\int_{S_2} g_1(s, t)\sigma(t)dt = -u_1(s). \tag{13.1.19}$$

Let us check that

$$g_1(s, t) = \frac{1}{4\pi|s - t|}\left(1 + O(|s - t|)\right) \quad \text{as } |s - t| \to 0. \tag{13.1.20}$$

Lemma 13.1.1. *Estimate* (13.1.20) *holds if* $dist(s, S_1) > 0$.

Proof. Let $g(x, y) = \frac{e^{ik|x-y|}}{4\pi|x-y|}$. By Green's formula, one has

$$g_1(x, y) = g(x, y) + \int_{S_1} g(x, t')\frac{\partial g_1(t', y)}{\partial N_{t'}}dt' \tag{13.1.21}$$

If $x = s$, $y = t$, and the distance from s to S_1 is positive, then the distance from t to S_1 is positive as $|t - s| \to 0$. Consequently, $|t' - y| \geq c > 0$, $|x - t'| \geq c > 0$, and the integral in (13.1.21) is a bounded function when $|s - t| \to 0$. Moreover, this function is infinitely differentiable with respect to s and t. The function $g(x, y)$ clearly has the form (13.1.20). Therefore, the right-hand side of (13.1.21) can be written as

$$g_1(x, y) = g(x, y)\left(1 + O(|x - y|)\right) = \frac{1}{4\pi|x - y|}\left(1 + O(|x - y|)\right) \tag{13.1.22}$$

Lemma 13.1.1 is proved. $\qquad\qquad\qquad\qquad\qquad\qquad\qquad\square$

If a is small, then, taking into account (13.1.20) and neglecting the error term $O(|s-t|)$, one gets the integral equation

$$\int_{S_2} \frac{\sigma(t)dt}{4\pi|s-t|} = -u_1(\xi), \tag{13.1.23}$$

where the terms of higher order of smallness as $a \to 0$ are neglected.

Equation (13.1.23) can be interpreted as an equation for the electrostatic charge distribution $\sigma(t)$ on the surface S_2 of a perfect conductor charged to a constant potential $U := u_1(\xi)$. The quantity $Q = \int_{S_2} \sigma(t)dt$ is the total charge on S_2. It is known that

$$Q = CU = -Cu_1(\xi), \tag{13.1.24}$$

where C is the electrical capacitance of the perfect conductor which has the shape of D_2. Therefore, the field v, scattered by both bodies D_1 and D_2, can be calculated analytically by formula (13.1.18) as follows:

$$v = \frac{e^{ik|x|}}{|x|}\left(A_1(\beta,\alpha) - Cu_1(\xi)\phi(\xi, -\beta, k)\right). \tag{13.1.25}$$

In this formula, v is known from the measurements at the large distance $|x|$ in the direction β, and one wants to calculate ξ and C from these measurements. The quantities $A_1(\beta,\alpha)$ and the functions u_1 and ϕ are known, $|x|$ and β are known, C and ξ should be found. Since $A_1(\beta,\alpha)$, v and x are known, one may assume that the quantity

$$Cu_1(\xi)\phi_1(\xi, -\beta, k) = A_1(\beta,\alpha) - ve^{-ik|x|}|x| \tag{13.1.26}$$

is known. Finally, the function $\phi_1(\xi, -\beta, k)$ has the following asymptotics for large $|\xi|$:

$$\phi_1(\xi, -\beta, k) = e^{-ik\beta\cdot\xi} + O\left(\frac{1}{|\xi|}\right), \quad |\xi| \to \infty. \tag{13.1.27}$$

Neglecting the small term $O\left(\frac{1}{|\xi|}\right)$, one wants to estimate C and ξ from the data (cf (13.1.26)):

$$f(\xi, C, \beta) = Cu_1(\xi)e^{-ik\beta\cdot\xi}. \tag{13.1.28}$$

Suppose that for β_1 and β_2 the values of f are measured and $u_1 \neq 0$. Then

$$\frac{f(\xi, C, \beta_1)}{f(\xi, C, \beta_2)} = e^{-ik(\beta_1 - \beta_2) \cdot \xi} \tag{13.1.29}$$

Denote $\beta_1 - \beta_2 := \theta$. Then

$$\ln \frac{f(\xi, C, \beta_1)}{f(\xi, C, \beta_2)} = -ik\theta \cdot \xi + 2\pi i n, \tag{13.1.30}$$

where n is an unknown integer. Differentiating formula (13.1.30) with respect to θ, one gets rid of the unknown n, because n is not changed when θ is changed a little. Thus, one gets

$$\xi = -\frac{1}{ik} \nabla_\theta \ln \frac{f(\xi, C, \beta_2 + \theta)}{f(\xi, C, \beta_2)}, \quad \theta = \beta_1 - \beta_2, \tag{13.1.31}$$

where β_2 is a constant here. Note that the right-hand side of (13.1.31) can be written as

$$\xi = \frac{i}{k} \frac{\nabla_\theta f(\xi, C, \beta_2 + \theta)}{f(\xi, C, \beta_2 + \theta)}. \tag{13.1.32}$$

If ξ is found, then formula (13.1.28) yields

$$C = \frac{f(\xi, C, \beta)}{u_1(\xi)} e^{ik\beta \cdot \xi}. \tag{13.1.33}$$

Since $C = O(a)$, the value of C can be considered as an estimate of the size of D_2.

In practice, one can measure the quantity f in (13.1.28) with some error $\delta > 0$. In this case, one has to differentiate stably the noisy data f_δ, $|f_\delta - f| \leq \delta$.

Methods for stable differentiation of noisy data are given by the author in Ramm (2007h).

13.2 Finding small subsurface inhomogeneities

In this section, the following problem is discussed. Suppose that a piece of metal or other material has small holes inside. One irradiates this object by acoustic waves, measures the scattered field on the surface of the object, and wants to find out if there exist small holes or cracks in it. If they exist, one wants to find their locations and their sizes. Similar questions can be asked about any small inhomogeneities which are located inside an object.

The other example deals with a problem in ultrasound mammography. Currently, X-rays are widely used for an early detection of cancer cells in a woman's breast. There is some probability for a woman to get a new cancer cell as a result of X-ray test. Therefore, there is an opinion that an ultrasound test is more appropriate for the purpose of detecting an early formation of cancer cells in a woman's breast. In the ultrasound mammography, one places a source of ultrasound waves on a woman's breast and measures the scattered field on the surface of the breast. Then one changes the position of the source and collects the measured data again, etc. From the measurements of the scattered field on the surface of the breast for various positions of the source and the receiver on the surface of the breast, one wants to find if there are some cancer cells inside the breast and if there are, what their locations are and what their sizes are.

The author has proved (Ramm (1992, 2005a)) the uniqueness of the solution to the geophysical inverse problem (IP), which covers the problems mentioned above. The precise formulation of IP is as follows. Let

$$\left[\nabla^2 + k^2 + k^2 v(x)\right] u = -\delta(x - y) \quad \text{in } \mathbb{R}^3, \qquad (13.2.1)$$

where $k > 0$ is the wave number, $v(x)$ is an inhomogeneity, and y is the position of the point source of the acoustic field $u = u(x, y)$. We assume that support of v is located in the half-space $x_3 < 0$ and is a bounded region, and that the surface $P = \{x : x_3 = 0\}$ is the surface on which $u(x, y)$ is measured, so $v(x)$ is the subsurface inhomogeneity. Finding v from the scattering data measured on the surface is an inverse problem of geophysics.

Inverse Problem of Geophysics (IPG):
Given $u(x, y)$ for all $x, y \in P$, find $v(x)$.

It is assumed that $v(x) \in L^2(\Omega)$, where $\overline{\Omega}$ is the support of v, a closure of a bounded domain. In many applications, $v(x) \in L^\infty(\Omega)$. This is so in both examples we mentioned above. Uniqueness of the solution to IP is proved by the author even in the case when the data $u(x, y)$ are known for $x \in P_1$ and $y \in P_2$, where $P_j \in P$, $j = 1, 2$, are arbitrary fixed open subsets of P, however small. However, the computational procedure for finding $v(x)$ from the data $u(x, y)$ is complicated. This procedure was developed and justified even in the case of noisy data, and error estimates were derived (see Ramm (2005a)).

In this section, a different procedure will be described, a "parameter estimation" procedure. This procedure is much easier to describe than the exact inversion procedure developed in Ramm (2005a).

Our basic assumption will be the assumption concerning the smallness of the components of the support of v. Namely, assume that

$$\text{supp}(v) = \bigcup_{m=1}^{M} B_m(\tilde{x}_m, r_m), \quad \sup_{x \in \mathbb{R}^3} |v(x)| \leq c_v, \qquad (13.2.2)$$

where B_m are balls centered at the points $\tilde{z}_m$ of radii r_m, $\tilde{z}_m \in \mathbb{R}^3_- := \{x : x_3 < 0\}$,

$$\sup_{1 \leq m \leq M} r_m = a, \quad ka \ll 1. \qquad (13.2.3)$$

Let us discuss the following inverse problem:

IP:

Given $u(x, y)$ for all $x, y \in P$, find M, $\tilde{z}_m$, r_m and $\tilde{v}_m := \int_{B_m} v(x)dx$.

Thus, our problem consists of finding a finite number of parameters. To explain our method for finding these parameters, let us introduce some notations.

Let

$$\{x_j, y_j\} := \xi_j, \quad 1 \leq j \leq J, \quad x_j, y_j \in P, \qquad (13.2.4)$$

be the pairs of the positions of the source and receiver at which the data $u(x_j, y_j)$ are measured,

$$g(x, y) := \frac{e^{ik|x-y|}}{4\pi|x - y|}, \quad k > 0 \text{ is fixed}, \qquad (13.2.5)$$

$$G_j(z) := G(\xi_j, z) := g(x_j, z)g(z, y_j), \qquad (13.2.6)$$

$$f_j := \frac{u(x_j, y_j) - g(x_j, y_j)}{k^2}, \qquad (13.2.7)$$

and

$$\Phi(z_1,\ldots,z_M,v_1,\ldots,v_N) := \sum_{j=1}^{J}\left|f_j - \sum_{m=1}^{M} G_j(z_m)v_m\right|^2. \qquad (13.2.8)$$

The method for solving *IP*, that is discussed in this section, consists of finding a global minimizer for the function (13.2.8). This minimizer gives an estimate of the positions z_m and the intensities v_m of the small inhomogeneities, as well as of their number M. The number M in this problem is assumed not large, say, $M \leq 20$. An estimate of M is a part of the procedure described in what follows. Our method is valid in the situation when the Born approximation is not applicable. The Born approximation consists of the following. Consider the exact integral equation for u:

$$u(x,y) = g(x,y) + k^2 \int_\Omega g(x,z)v(z)u(z,y)dz. \qquad (13.2.9)$$

If $u(z,y)$ can be replaced by $g(x,y)$ with a small relative error, then $u(x,y)$ can be approximated by the expression

$$u(x,y) \approx g(x,y) + k^2 \int_\Omega g(x,z)v(z)g(z,y)dz \qquad (13.2.10)$$

linear in v. This approximation reduces a strongly nonlinear inverse problem of finding $v(z)$ from the data $u(x,y)$, $x,y \in P$, to a linear problem of finding $v(z)$ from the following linear equation:

$$\int_\Omega g(x,z)g(z,y)v(z)dz = [u(x,y) - g(x,y)]k^{-2}. \qquad (13.2.11)$$

However, this approximation requires the smallness of v and is not applicable, for example, when $v(z) = \delta(z)$, where $\delta(z)$ is the delta-function. In this case, equation (13.2.9) yields

$$u(x,y) = g(x,y) + k^2 g(x,0)u(0,y), \qquad (13.2.12)$$

and one cannot find $u(0,y)$ by taking $x = 0$ in (13.2.12) because $g(0,0) = \infty$. In our case, the inhomogeneity

$$v(x) = \sum_{m=1}^{M} \tilde{v}_m \delta(x - \tilde{z}_m), \qquad (13.2.13)$$

so the Born approximation is not applicable in its usual form.

We write the exact equation (13.2.9) as

$$u(x,y) = g(x,y) + k^2 \sum_{m=1}^{M} \int_{B_m} g(x,z)v(z)u(z,y)dz, \qquad (13.2.14)$$

and approximate the term $\int_{B_m} g(x,z)v(z)u(z,y)dz$ as follows:

$$\int_{B_m} g(x,z)v(z)u(z,y)dz = g(x,\tilde{z}_m)\tilde{v}_m u(\tilde{z}_m,y). \qquad (13.2.15)$$

We assume that the distance from $z \in B_m$ to the plane P is not less than d and use the following estimates:

$$|g(x,z)| \leq \frac{1}{4\pi d}, \quad x \in P, \quad z \in B_m, \quad 1 \leq m \leq M, \qquad (13.2.16)$$

$$|u(z,y)| \leq \frac{c}{d}, \quad x \in P, \quad z \in B_m, \quad 1 \leq m \leq M, \qquad (13.2.17)$$

where $c > 0$ is some constant. Consequently, from (13.2.14) one concludes that

$$\left| k^2 \sum_{m=1}^{M} \int_{B_m} g(x,z)v(z)u(z,y)dz \right| \leq \frac{k^2 M c_v c a^3}{d^2}. \qquad (13.2.18)$$

This term is much less than $|g(x,y)|$ if

$$M c_v c \frac{k^2 a^3}{d^2} \ll \frac{1}{4\pi|x-y|}. \qquad (13.2.19)$$

The constant c in (13.2.17) and (13.2.19) is of the order $O(1)$, $M \leq 20$, and c_v is of order $O(1)$. Therefore, (13.2.19) holds if

$$\frac{k^2 a^3}{d^2} \ll \frac{1}{|x-y|}. \qquad (13.2.20)$$

Condition (13.2.20) holds if

$$\frac{k^2 a^3}{d} \ll \frac{d}{|x-y|}. \qquad (13.2.21)$$

This condition is satisfied if, for example, $ka \ll 1$, $\frac{a}{d} = O\left(\frac{d}{|x-y|}\right)$.

If this condition holds, then $u(x, y)$ can be well approximated on P by $g(x, y)$ and equation (13.2.14) implies

$$[u(x, y) - g(x, y)] \, k^{-2} = \sum_{m=1}^{M} g(x, \tilde{z}_m) \tilde{v}_m g(\tilde{z}_m, y). \qquad (13.2.22)$$

Thus the quantity

$$\Phi := \sum_{j=1}^{J} \left| f_j - \sum_{m=1}^{M} G(\xi_m, z_m) v_m \right|^2 \qquad (13.2.23)$$

should attain its global minimum when $v_m = \tilde{v}_m$ and $z_m = \tilde{z}_m$. This justifies our method. How does one estimate M and compute the minimizer of Φ? A numerical procedure for doing this is described in Gutman and Ramm (2000) and illustrated by numerical examples.

13.3   Inverse radiomeasurements problem

In this section, we solve the following problem. In many applications it is important to know accurately the complicated electromagnetic field distribution in the aperture of an antenna. This information allows one to answer nearly all practical questions concerning this antenna. Suppose a mirror antenna in radio waves diapason $\lambda = 3$ cm is considered, a small probe, say of the characteristic size $a = 1$ mm, is moved in the aperture of the antenna by a mechanical device transparent for waves at the wavelength $\lambda = 3$ cm and one can measure the field E, H scattered by the probe for various positions of this probe in the aperture of the antenna. Finally, we assume that the shape of the probe and its parameters ϵ, μ, σ are known, where ϵ is its permittivity, μ is its permeability, and σ is its conductivity.

The inverse radiomeasurements problem can be formulated as follows:

IRP:
Given the scattered field E, H in the far zone from the small probe, calculate the initial field E_0, H_0 at the point where the probe was located before this probe was put at this point of the aperture of the antenna.

We prove that this is possible and give an analytical solution to this inverse problem. Let us assume for simplicity that $\mu = \mu_0$ and $\sigma = 0$ in the probe where μ_0 is the magnetic permeability of the air, and ϵ is the electrical constant of the probe, while ϵ_0 is this constant in the air. Then the electric field, scattered by the probe in the direction N, is given by the following formula (see Landau and Lifshitz (1984) and Ramm (2005b)):

$$E = \frac{k^2}{4\pi\epsilon_0}[N, [P, N]], \qquad (13.3.1)$$

where P is the dipole moment induced on the small probe by the original field E_0 and we neglect the magnetic dipole radiation. This is possible, for example, if the skin-depth is much larger than the size of the probe. Since we assumed that $ka \ll 1$, it follows that the induced dipole moment can be calculated by solving the electrostatic problem of putting the probe into a constant (homogeneous) static field E_0 calculated at the point where the probe was located. The solution to this static problem is given by the following formula:

$$P_j = \sum_{m=1}^{3} \alpha_{jm}\epsilon_0 V E_{0m}, \quad 1 \leq j \leq 3, \qquad (13.3.2)$$

where V is the volume of the probe, which is known, and α_{jm} is the polarizability tensor of the probe, which depends on the parameter $\gamma := \frac{\epsilon-\epsilon_0}{\epsilon+\epsilon_0}$. Analytical formulas which allow one to calculate tensor α_{jm} for a body of an arbitrary shape are derived in Ramm (2005b). Therefore, we assume α_{jm} to be known.

Denote

$$b := \frac{k^2}{4\pi\epsilon_0}, \qquad (13.3.3)$$

and write formula (13.3.1) for the directions $N^{(j)}$ as

$$E = b\left(P - N^{(j)}P \cdot N^{(j)}\right), \quad j = 1, 2, 3, \qquad (13.3.4)$$

where over the repeated indices j there is no summation, $A \cdot B$ denotes the inner product of two vectors, and $N^{(j)}$ are orthogonal to each other's unit vectors. One has

$$bP = E + N^{(j)}P \cdot N^{(j)},$$

where there is no summation over j. Let $j = 2$ in this equation, and take the scalar product of this equation with vector $N^{(1)}$. The result is $bP \cdot N^{(1)} = E \cdot N^{(1)}$. In a similar way, one derives the following three equations:

$$bP \cdot N^{(j)} = E \cdot N^{(j)}, \quad j = 1, 2, 3. \tag{13.3.5}$$

Consequently, one can calculate vector P from the scattered field E:

$$P = b^{-1} \sum_{j=1}^{3} E \cdot N^{(j)} N^{(j)}. \tag{13.3.6}$$

If P is found by formula (13.3.6), then E_0 can be found from the following LAS:

$$\sum_{m=1}^{3} \alpha_{jm} \epsilon_0 V E_{0m} = P_j, \quad 1 \le j \le 3. \tag{13.3.7}$$

The matrix α_{jm} of the LAS (13.3.7) is positive definite, see Landau and Lifshitz (1984) and Ramm (2005b), so this system is uniquely solvable for E_0. Therefore, IRP is solved analytically. The important ingredients in our solution of this problem are explicit analytical formulas allowing one to calculate the matrix (tensor) α_{jm} with a desired accuracy for a probe of an arbitrary shape. These formulas are derived in Ramm (2005b) and originally in Ramm (1969b) and Ramm (1970).

13.4 Summary of the results

The basic results of this chapter include a method for finding the position of a small body from the scattering data described in Section 13.1, see formulas (13.1.32), (13.1.33); a method for finding several small subsurface inhomogeneities from the scattering data collected on the surface; a solution of the inverse radiomeasurements problem.

Appendix A

A.1 Banach and Hilbert spaces

A linear vector space X is a Banach space if with every element $u \in X$ a number $||u||$ is associated, called the norm of this element, and this number has the following properties

$$||u|| \geq 0, \quad ||u|| = 0 \Leftrightarrow u = 0, \tag{A.1.1}$$

$$||\lambda u|| = |\lambda| ||u||, \quad \forall \lambda \in \mathbb{C}, \tag{A.1.2}$$

$$||u + v|| \leq ||u|| + ||v||, \tag{A.1.3}$$

and X is complete with respect to this norm. By $\mathbb{C}$, the set of complex numbers is denoted. Completeness with respect to the norm $||u||$ means that every Cauchy sequence converges to an element $u \in X$, namely, if $||u_n - u_m|| \to 0$ when $n, m \to \infty$, then there exists $u \in X$ such that $u = \lim_{n\to\infty} u_n$, that is $\lim_{n\to\infty} ||u_n - u|| = 0$.

Examples of Banach spaces are numerous. The space $C(D)$ of continuous functions in a bounded domain $D \subset \mathbb{R}^n$ with the norm $||u|| = \max_{x\in D} |u(x)|$ is a Banach space. The space $L^p(D)$ of square integrable functions in D with the norm $||u||_{L^p(D)} := \left(\int_D |u(x)|^p dx\right)^{1/p}, p \geq 1$, is a Banach space. This space is also a Hilbert space H if $p = 2$.

A linear vector space is a Hilbert space if with every ordered pair of its elements u, v, a form (u, v) is associated, and this form has the following properties:

$$(u, v) = \overline{(v, u)}, \tag{A.1.4}$$

$$(\lambda u + \mu v, w) = \lambda(u, w) + \mu(v, w), \quad \forall \lambda, \mu \in \mathbb{C}, \tag{A.1.5}$$

$$(u, u) \geq 0, \quad (u, u) = 0 \Leftrightarrow u = 0, \tag{A.1.6}$$

and H is complete with respect to the norm $||u|| := (u, u)^{1/2}$. From this definition, one derives the Cauchy inequality

$$|(u, v)| \leq ||u|| \, ||v||, \tag{A.1.7}$$

where the equality sign is attained if and only if u and v are linearly dependent.

Examples of Hilbert spaces include the Sobolev spaces H^m of functions which are square integrable in a domain D together will all their derivatives $D^l u$ of the order $l \leq m$, and

$$(u, v)_m = \left(\sum_{l=0}^{m} (D^l u, D^l v) \right)^{1/2}. \tag{A.1.8}$$

The Sobolev space $W^{m,p}(D)$ is a Banach space with the norm

$$||u||_{W^{m,p}(D)} := \sum_{l=0}^{m} ||D^l u||_{L^p(D)}. \tag{A.1.9}$$

By $D^l u$, an arbitrary derivative of the order l is meant, as follows:

$$\frac{\partial^l u}{\partial x^{l_1} \partial x^{l_2} \ldots \partial x^{l_n}}, \quad l = \sum_{j=1}^{n} l_j, \quad l_j \geq 0.$$

A result often useful in applications is the embedding theorem, which is formulated as follows (see Gilbarg and Trudinger (1983)).

Theorem A.1.1. *Let $D \subset \mathbb{R}^n$ be a bounded somain. If $mp < n$, then the space $W^{m,p}(D)$ is continuously embedded in $L^{p^*}(D)$, $p^* := \frac{np}{n-mp}$, and compactly imbedded in $L^q(D)$ for $q < p^*$.*

If $0 \leq j < m - \frac{n}{p} < j+1$, then $W^{m,p}$ is continuously imbedded in $C^{j,\alpha}(\overline{D}), \alpha := m - \frac{n}{p} - j$, and compactly imbedded in $C^{j,\beta}(\overline{D})$, for $\beta < \alpha$.

Let us define the imbedding operator. Suppose that $X \subset Y$, where X and Y are Banach spaces, $||u||_X \geq ||u||_Y, \forall u \in X$. Then each element of X can be considered as an element of Y. The operator $i : X \to Y$ is called the imbedding operator, $||iu||_Y \leq ||u||_X, \forall u \in X$. Therefore, $||i||_{X \to Y} \leq 1$. If $i : X \to Y$ is compact, then any bounded in X sequence of elements contains a subsequence convergent in Y. Theorem A.1.1 gives examples of the compact imbedding operators.

A.2   A result from perturbation theory

In this section, we prove the following result (see, for example, Kato (1984)):

The set of bounded invertible linear operators is open:

If A is a linear bounded operator and A^{-1} is a bounded operator, then B^{-1} is a bounded operator if $||B - A||$ is sufficiently small.

To prove this result one argues as follows:

$$B = A + B - A = A[I + A^{-1}(B - A)].$$

It is easy to see that the operator $(I + T)^{-1}$ exists and is bounded if $||T|| < 1$. Indeed,

$$(I + T)^{-1} = \sum_{j=1}^{\infty} (-1)^j T^j,$$

where the series converges in the norm of operators and is majorized by the series

$$\sum_{j=1}^{\infty} ||T^j|| = (1 - ||T||)^{-1}.$$

Therefore,

$$||(I + T)^{-1}|| \leq \frac{1}{1 - ||T||}.$$

A.3　The Fredholm alternative

The aim of this section is to prove the Fredholm alternative and to give a characterization of the class of Fredholm operators in a very simple way, by a reduction of the operator equation with a Fredholm operator to a linear algebraic system in a finite-dimensional space.

The Fredholm alternative is a classical well-known result whose proof for linear equations of the form $(I + T)u = f$, where T is a compact operator in a Banach space, can be found in most texts on functional analysis, of which we mention just Kato (1984) and Kantorovich and Akilov (1982). A characterization of the set of Fredholm operators is in Kantorovich and Akilov (1982), p. 500, but it is not given in the majority of the books. The proofs in many books follow the classical Riesz argument used in developing the Riesz–Fredholm theory.

In our book, a Fredholm-type operator is an operator of the form $A = B + T$, where $B : X_1 \to X_2$ is a linear isomorphism of a Banach space X_1 onto a Banach space X_2, and $T : X_1 \to X_2$ is a linear compact operator. The index of A is defined as $indA := \dim N(A) - \dim N(A^*)$. *In our book, the index of Fredholm-type operators is equal to zero.* In the literature (see Kato (1984)), the Fredholm operators are defined in a more general way and they may have non-zero index. Example of such operators are singular integral operators.

Our aim is to give a short and simple proof of the Fredholm alternative and of a characterization of the class of Fredholm operators. We give the argument for the case of the Hilbert space, but the proof is quite easy to adjust for the case of the Banach space.

The idea is to reduce the problem to the one for linear algebraic systems in finite-dimensional case, for which the Fredholm alternative is a basic fact: in a finite-dimensional space, $\mathbb{R}^N$ property (A.3.4) in the Definition A.3.1 of Fredholm operators is a consequence of the closedness of any finite-dimensional linear subspace, since $R(A)$ is such a subspace in $\mathbb{R}^N$, while property (A.3.3) is a consequence of the simple formulas $r(A) = r(A^*)$ and $n(A) = N - r(A)$, valid for matrices, where $r(A)$ is the rank of A and $n(A)$ is the dimension of the null-space of A.

Throughout the book, A is a linear bounded operator, A^* is its adjoint, and $N(A)$ and $R(A)$ are the null-space and the range of A, respectively.

Recall that an operator F with $dim\, R(F) < \infty$ is called a *finite-rank operator*, its rank is $n := dim\, R(F)$.

We call a linear bounded operator B on H an *isomorphism* if it is a bicontinuous injection of H onto H, that is, B^{-1} is defined on all of H and is bounded.

If $e_j, 1 \leq j \leq n$, is an orthonormal basis of $R(F)$, then $Fu = \sum_{j=1}^{n}(Fu, e_j)e_j$, so

$$Fu = \sum_{j=1}^{n}(u, F^*e_j)e_j, \qquad (A.3.1)$$

and

$$F^*u = \sum_{j=1}^{n}(u, e_j)F^*e_j, \qquad (A.3.2)$$

where (u, v) is the inner product in H.

Definition A.3.1. An operator A is called Fredholm (or Fredholm-type) if and only if

$$dim\, N(A) = dim\, N(A^*) := n < \infty, \qquad (A.3.3)$$

and

$$R(A) = \overline{R(A)}, \quad R(A^*) = \overline{R(A^*)}, \qquad (A.3.4)$$

where the overline stands for the closure.

Recall that

$$H = \overline{R(A)} \oplus N(A^*), \quad H = \overline{R(A^*)} \oplus N(A), \qquad (A.3.5)$$

for any linear densely defined (i.e., having a domain of definition dense in H) operator A, not necessarily bounded. The sign $\oplus$ stands for the orthogonal sum.

For a Fredholm operator A, one has

$$H = R(A) \oplus N(A^*), \quad H = R(A^*) \oplus N(A). \qquad (A.3.6)$$

Consider the following equations:

$$Au = f, \qquad\qquad (A.3.7)$$

$$Au_0 = 0, \qquad\qquad (A.3.8)$$

$$A^*v = g, \qquad\qquad (A.3.9)$$

$$A^*v_0 = 0. \qquad\qquad (A.3.10)$$

Let us formulate the Fredholm alternative:

Theorem A.3.1. *If B is an isomorphism and F is a finite rank operator, then $A = B + F$ is Fredholm.*

For any Fredholm operator A, the following (Fredholm) alternative holds:

(1) *Either equation (A.3.8) has only the trivial solution $u_0 = 0$, and then equation (A.3.10) has only the trivial solution, and equations (A.3.7) and (A.3.9) are uniquely solvable for any right-hand sides f and g,*

   or

(2) *Equation (A.3.8) has exactly $n > 0$ linearly independent solutions $\{\phi_j\}, 1 \le j \le n$, and then equation (A.3.10) has also n linearly independent solutions $\{\psi_j\}, 1 \le j \le n$, equations (A.3.7) and (A.3.9) are solvable if and only if $(f, \psi_j) = 0, 1 \le j \le n$, and correspondingly $(g, \phi_j) = 0, 1 \le j \le n$. If they are solvable, their solutions are not unique and their general solutions are, respectively: $u = u_p + \sum_{j=1}^{n} a_j \phi_j$, and $v = v_p + \sum_{j=1}^{n} b_j \psi_j$, where a_j and b_j are arbitrary constants, and u_p and v_p are some particular solutions to equations (A.3.7) and (A.3.9), respectively.*

The number n in the previous theorem is not related to the number n in the definition of the finite-rank operator.

Let us give a characterization of the class of Fredholm operators, that is, a necessary and sufficient condition for A to be Fredholm.

Theorem A.3.2. *A linear bounded operator A is Fredholm if and only if $A = B + F$, where B is an isomorphism and F has finite rank.*

Let us prove these theorems.

Proof of Theorem A.3.2. From the proof of Theorem A.3.1 that follows, we see that if $A = B + F$, where B is an isomorphism and F has finite rank, then A is Fredholm. To prove the converse, choose some orthonormal bases $\{\phi_j\}$ and $\{\psi_j\}$, in $N(A)$ and $N(A^*)$, respectively, using assumption (A.3.3). Define

$$Bu := Au - \sum_{j=1}^{n} (u, \phi_j)\psi_j := Au - Fu. \tag{A.3.11}$$

Clearly, F has finite rank, and $A = B + F$. Let us prove that B is an isomorphism. If this is done, then Theorem A.3.2 is proved.

We need to prove that $N(B) = \{0\}$ and $R(B) = H$. It is known (Banach's theorem, see Kantorovich and Akilov (1982)), that if B is a linear injection and $R(B) = H$, then B^{-1} is a bounded operator, so B is an isomorphism.

Suppose $Bu = 0$. Then $Au = 0$ (so that $u \in N(A)$), and $Fu = 0$ (because, according to (A.3.6), Au is orthogonal to Fu). Since $\{\psi_j\}, 1 \leq j \leq n$, is a linearly independent system, the equation $Fu = 0$ implies $(u, \phi_j) = 0$ for all $1 \leq j \leq n$, that is, u is orthogonal to $N(A)$. If $u \in N(A)$ and at the same time it is orthogonal to $N(A)$, then $u = 0$. So, $N(B) = \{0\}$.

Let us now prove that $R(B) = H$:

Take an arbitrary $f \in H$ and, using (A.3.6), represent it as $f = f_1 + f_2$ where $f_1 \in R(A)$ and $f_2 \in N(A^*)$ are orthogonal. Thus, there is a $u_p \in H$ and some constants c_j such that $f = Au_p + \sum_1^n c_j \psi_j$. We choose u_p orthogonal to $N(A)$. This is clearly possible.

We claim that $Bu = f$, where $u := u_p - \sum_1^n c_j \phi_j$. Indeed, using the orthonormality of the system ϕ_j, $1 \leq j \leq n$, one gets $Bu = Au_p + \sum_1^n c_j \psi_j = f$.

Thus, we have proved that $R(B) = H$. $\square$

We now prove Theorem A.3.2.

Proof of Theorem A.3.1. If A is a Fredholm operator, then the statements (1) and (2) of Theorem A.3.1 are equivalent to (A.3.3) and (A.3.4), since (A.3.6) follows from (A.3.4).

Let us prove that if $A = B + F$, where B is an isomorphism and F has finite-rank, then A is a Fredholm operator. Both properties (A.3.3) and (A.3.4) are known for operators in finite-dimensional spaces. Therefore, to prove that A is Fredholm it is sufficient to prove that equations (A.3.7) and (A.3.9) are equivalent to linear algebraic systems in a finite-dimensional space.

Let us prove this equivalence. We start with equation (A.3.7), denote $Bu := w$, and get an equation

$$w + Tw = f, \tag{A.3.12}$$

that is equivalent to (A.3.7). Here, $T := FB^{-1}$ is a finite rank operator that has the same rank n as F because B is an isomorphism. Equation (A.3.12) is equivalent to (A.3.7): each solution to (A.3.7) is in one-to-one correspondence with a solution of (A.3.12) since B is an isomorphism. In particular, the dimensions of the null-spaces $N(A)$ and $N(I + T)$ are equal, $R(A) = R(I + T)$, and $R(I + T)$ is closed. The last claim is a consequence of the Fredholm alternative for finite-dimensional linear equations, but we give an independent proof of the closedness of $R(A)$ at the end of the book.

Since T is a finite rank operator, the dimension of $N(I + T)$ is finite and is not greater than the rank of T. Indeed, if $u = -Tu$ and T has finite rank n, then $Tu = \sum_{j=1}^{n}(Tu, e_j)e_j$, where $\{e_j\}_{1 \le j \le n}$, is an orthonormal basis of $R(T)$, and $u = -\sum_{j=1}^{n}(u, T^*e_j)e_j$, so that u belongs to a subspace of dimension $n = r(T)$.

Since A and A^* enter symmetrically in the statement of Theorem A.3.1, it is sufficient to prove (A.3.3) and (A.3.4) for A and check that the dimensions of $N(A)$ and $N(A^*)$ are equal.

To prove (A.3.3) and (A.3.4), let us reduce (A.3.9) to an equivalent equation of the form

$$v + T^*v = h, \tag{A.3.13}$$

where $T^* := B^{*-1}F^*$ is the adjoint to T, and

$$h := B^{*-1}g. \tag{A.3.14}$$

Since B is an isomorphism, one has $(B^{-1})^* = (B^*)^{-1}$. Applying B^{*-1} to equation (A.3.9), one gets an equivalent equation (A.3.13) and T^* is a finite-rank operator of the same rank n as T.

The last claim is easy to prove: if $\{e_j\}_{1 \le j \le n}$ is a basis in $R(T)$, then $Tu = \sum_{j=1}^{n}(Tu, e_j)e_j$, and $T^*u = \sum_{j=1}^{n}(u, e_j)T^*e_j$, so $r(T^*) \le r(T)$. By symmetry, one has $r(T) \le r(T^*)$, and the claim is proved.

Writing explicitly the linear algebraic systems, equivalent to the equations (A.3.12) and (A.3.13), one sees that the matrices of these systems are adjoint. The system equivalent to equation (A.3.12) is

$$c_i + \sum_1^n t_{ij} c_j = f_i, \qquad (A.3.15)$$

where

$$t_{ij} := (e_j, T^* e_i), \quad c_j := (w, T^* e_j), \quad f_i := (f, T^* e_i),$$

and the one equivalent to (A.3.13) is

$$\xi_i + \sum_1^n t_{ij}^* \xi_j = h_i, \qquad (A.3.16)$$

where

$$t_{ij}^* = (T^* e_j, e_i), \quad \xi_j := (v, e_j), \quad h_i := (h, e_i),$$

and t_{ij}^* is the matrix adjoint to t_{ij}. For linear algebraic systems (A.3.15) and (A.3.16), the Fredholm alternative is a well-known elementary result. These systems are equivalent to equations (A.3.7) and (A.3.9), respectively. Therefore, the Fredholm alternative holds for equations (A.3.7) and (A.3.9), so that properties (A.3.3) and (A.3.4) are proved. Thus, Theorem A.3.2 is proved. $\qquad\square$

In conclusion, let us explain in detail why equations (A.3.12) and (A.3.15) are equivalent in the following sense:

Every solution to (A.3.12) *generates a solution to* (A.3.15) *and vice versa.*

It is clear that (A.3.12) implies (A.3.15): just take the inner product of (A.3.12) with $T^* e_j$ and get (A.3.15). So, each solution to (A.3.12) generates a solution to (A.3.15).

We claim that each solution to (A.3.15) generates a solution to (A.3.12). Indeed, let c_j solve (A.3.15). Define $w := f - \sum_1^n c_j e_j$. Then $Tw = Tf - \sum_{j=1}^n c_j T e_j = \sum_{i=1}^n [(Tf, e_i) e_i - \sum_{j=1}^n c_j (T e_j, e_i) e_i] = \sum_{i=1}^n c_i e_i = f - w$. Here we use (A.3.15) and take into account that $(Tf, e_i) = f_i$ and $(T e_j, e_i) = t_{ij}$. Thus, the element $w := f - \sum_1^n c_j e_j$ solves (A.3.12), as claimed.

It is easy to check that if $\{w_1, \ldots, w_k\}$ are k linearly independent solutions to the homogeneous version of equation (A.3.12), then the corresponding k solutions $\{c_{1m}, \ldots, c_{nm}\}_{1 \leq m \leq k}$ of the homogeneous version of the system (A.3.15) are also linearly independent, and vice versa.

Let us give an independent proof of property (A.3.4):

$R(A)$ *is closed if* $A = B + F$, *where* B *is an isomorphism and* F *is a finite rank operator.*

Since $A = (I + T)B$ and B is an isomorphism, it is sufficient to prove that $R(I + T)$ is closed if T has finite rank.

Let $u_j + Tu_j := f_j \to f$ as $j \to \infty$. Without loss of generality choose u_j orthogonal to $N(I + T)$. We want to prove that there exists a u such that

$$(I + T)u = f.$$

Suppose first that $\sup_{1 \leq j < \infty} \|u_j\| < \infty$, where $\| \cdot \|$ denotes the norm in H. Since T is a finite-rank operator, Tu_j converges in H for some subsequence, which is denoted by u_j again. Recall that in finite-dimensional spaces bounded sets are precompact. This implies that $u_j = f_j - Tu_j$ converges in H to an element u. Passing to the limit, one gets $(I + T)u = f$. To complete the proof, let us establish that $\sup_j \|u_j\| < \infty$. Assuming that this is false, one can choose a subsequence, denoted by u_j again, such that $\|u_j\| > j$. Let $z_j := u_j / \|u_j\|$. Then $\|z_j\| = 1$, z_j is orthogonal to $N(I + T)$, and $z_j + Tz_j = f_j / \|u_j\| \to 0$. As before, it follows that $z_j \to z$ in H, and passing to the limit in the equation for z_j one gets $z + Tz = 0$. Since z is orthogonal to $N(I + T)$, it follows that $z = 0$. This is a contradiction since $\|z\| = \lim_{j \to \infty} \|z_j\| = 1$. This contradiction proves the desired estimate and the proof is completed. $\qquad \square$

This proof is valid for any compact linear operator T. If T is a finite-rank operator, then the closedness of $R(I + T)$ follows from a simple observation: finite-dimensional linear spaces are closed.

Appendix B: Many-Body Wave Scattering Problems for Small Scatterers and Creating Materials with a Desired Refraction Coefficient*

Alexander G. Ramm

Department of Mathematics
Kansas State University, Manhattan, KS 66506-2602, USA
ramm@math.ksu.edu

Abstract

Formulas are derived for solutions of many-body wave scattering problems by small impedance particles embedded in an inhomogeneous medium. The limiting case is considered when the size a of small particles tends to zero while their number tends to infinity at a suitable rate. Equations for the limiting effective (self-consistent) field in the medium are derived. The theory is based on a study of integral equations and asymptotic of their solutions as $a \to 0$. The case of wave scattering by many small particles embedded in an inhomogeneous medium is also studied. Applications of this theory to creating materials with a desired refraction coefficient is given. A recipe is given for creating such materials by embedding into a given material many small impedance particles with prescribed boundary impedances.

---

*This chapter is published in *Mathematical Analysis and Applications*, Wiley, Hoboken, NJ, Chapter 3, pp. 57–76 (2018).

Keywords: wave scattering by many small bodies; materials science; smart materials.

1   Introduction

In this chapter, we discuss a method for creating materials with a desired refraction coefficients. This method is proposed and developed by the author and is based on a series of his papers and on his monograph [31]. The author thinks that these results may be new for materials science people although the results were published in mathematical and mathematical physics Journals. This is the basic reason for publishing this paper in a book possibly useful for materials science.

Parts of this chapter are taken verbatim from the paper [30]. The author thanks Springer-Verlag for permission to use parts of this paper verbatim.

There is a large literature on wave scattering by small bodies, starting from Rayleigh's work (1871), [35], [5], [3]. For the problem of wave scattering by one body, an analytical solution was found only for the bodies of special shapes, for example, for balls and ellipsoids. If the scatterer is small then the scattered field can be calculated analytically for bodies of arbitrary shapes, see [8], [30] and [31] where this theory is presented.

The many-body wave scattering problem was discussed in the literature, mostly numerically, in the cases when the number of scatterers is small or the influence on a particular particle of the waves scattered by other particles is negligible. This corresponds to the case when the distance d between neighboring particles is much larger than the wavelength λ, and the characteristic size a of a small body (particle) is much smaller than λ, that is, $d \gg \lambda$ and $a \ll \lambda$. By $k = \frac{2\pi}{\lambda}$ the wave number is denoted.

In this chapter, the much more difficult case is considered, when $a \ll d \ll \lambda$. In this case the influence of the scattered field on a particular particle is essential, that is, *multiple scattering effects are essential.*

The derivations of the results, presented in this chapter, are rigorous. They are taken from the earlier papers of the author, cited in the list of references. Many formulas and arguments are taken from these papers, especially from paper [30]. Large parts of this paper are taken verbatim, and monograph [31] is also used essentially. We do not discuss in this paper electromagnetic wave scattering by small bodies (particles). A detailed discussion of electromagnetic wave scattering by small perfectly conducting and impedance particles of an *arbitrary shape* is given in [34] and in [31], see also [8].

A *physically novel* point in our theory is the following one:

While in the classical theory of wave scattering by small body of characteristic size a (for example, in Rayleigh's theory) the scattering amplitude is $O(a^3)$ as $a \to 0$, in our theory for a small impedance particle the scattering amplitude is *much larger*: it is of the order $O(a^{2-\kappa})$, where $a \to 0$ and $\kappa \in (0, 1]$ is a parameter (see the text below formula (22) in this paper).

Can this result be used in technology?
The practical applications of the theory, presented in this paper, are immediate provided that the important practical problem of preparing small particles with the prescribed boundary impedance is solved.

The author thinks that an impedance boundary condition (condition (7) below) must be physically (experimentally) realizable if this condition guarantees the uniqueness of the solution to the corresponding boundary problem. The impedance boundary condition (7) guarantees the uniqueness of the solution to the scattering boundary problems (1)–(4) provided that $\mathrm{Im}\zeta_1 \leq 0$.

Therefore there should exist a practical (experimental) method for producing small particles with any boundary impedance ζ_1 satisfying the inequality $\mathrm{Im}\zeta_1 \leq 0$.

The author asks the materials science specialists to contact him if they are aware of a method for practical (experimental) preparing (producing) small particles with the prescribed boundary impedance.

The materials science researchers are not familiar with the author's papers on creating materials with a desired refraction coefficient because the author's theory was presented in the Journals which are not popular among materials science researchers.

Although my results were presented in many of my earlier publications, cited in references, I hope that *they will be not only new but practically useful for materials science researchers.*

The basic results of this section consist of:

(i) Derivation of analytic formulas for the scattering amplitude for the wave scattering problem by one small ($ka \ll 1$) impedance body *of an arbitrary shape*

(ii) Solution to *many-body wave scattering problem* by small particles, embedded in an inhomogeneous medium, under the assumptions $a \ll d \ll \lambda$, where d is the minimal distance between neighboring particles;

(iii) Derivation of the equations for the limiting effective (self-consistent) field in an inhomogeneous medium in which many small particles are embedded, when $a \to 0$ and the number $M = M(a)$ of the small particles tends to infinity at an appropriate rate;

(iv) Derivation of linear algebraic systems for solving many-body wave scattering problems. These systems are not obtained in the standard way from boundary integral equations; they have physical meaning and give an efficient numerical method for solving many-body wave scattering problems in the case of small scatterers. In [32] for the first time the many-body wave scattering problems were solved for billions of particles. This was not feasible earlier;

(v) Application of our results to creating materials with a desired refraction coefficient.

The order of the error estimates as $a \to 0$ is obtained. Our presentation follows very closely that in [30], but it is essentially self-contained. Our methods give powerful numerical methods for solving many-body wave scattering problems in the case when the scatterers are small but multiple scattering effects are essential (see [1], [2], [32]). In [32] the scattering problem is solved numerically for 10^{10} particles apparently for the first time.

In Sections 1–4 wave scattering by small impedance bodies is developed.

Let us formulate the wave scattering problems we deal with. First, let us consider a one-body scattering problem. Let D_1 be a bounded domain in $\mathbb{R}^3$ with a sufficiently smooth boundary S_1. The scattering problem consists of finding the solution to the problem:

$$(\nabla^2 + k^2)u = 0 \quad \text{in } D_1' := \mathbb{R}^3 \setminus D_1, \tag{1}$$

$$\Gamma u = 0 \quad \text{on } S_1, \tag{2}$$

$$u = u_0 + v, \tag{3}$$

where

$$u_0 = e^{ik\alpha \cdot x}, \quad \alpha \in S^2, \tag{4}$$

S^2 is the unit sphere in $\mathbb{R}^3$, u_0 is the incident field, v is the scattered field satisfying the radiation condition

$$v_r - ikv = o\left(\frac{1}{r}\right), \quad r := |x| \to \infty, \ v_r := \frac{\partial v}{\partial r}, \tag{5}$$

Γu is the boundary condition (bc) of one of the following types

$$\Gamma u = \Gamma_1 u = u \quad \text{(Dirichlet bc)}, \tag{6}$$

$$\Gamma u = \Gamma_2 u = u_N - \zeta_1 u, \quad \text{Im}\zeta_1 \le 0, \text{ (impedance bc)}, \tag{7}$$

where ζ_1 is a constant, N is the unit normal to S_1, pointing out of D_1, and

$$\Gamma u = \Gamma_3 u = u_N, \quad \text{(Neumann bc)}. \tag{8}$$

It is well known (see, e.g., [7]) that problems (1)–(3) have a unique solution. We now assume that

$$a := 0.5 \, \text{diam} D_1, \quad ka \ll 1, \tag{9}$$

which is the "smallness assumption" equivalent to $a \ll \lambda$, where λ is the wave length. We look for the solution to problems (1)–(3) of the form

$$u(x) = u_0(x) + \int_{S_1} g(x,t)\sigma_1(t)dt, \quad g(x,y) := \frac{e^{ik|x-y|}}{4\pi|x-y|}, \tag{10}$$

where dt is the element of the surface area of S_1. One can prove that the unique solution to the scattering problems (1)–(3) with any of the boundary conditions (6)–(8) can be found in the form (10), and the function σ_1 in equation (10) is uniquely defined from the boundary condition (2). The scattering amplitude $A(\beta, \alpha) = A(\beta, \alpha, k)$ is defined by the formula

$$v = \frac{e^{ikr}}{r} A(\beta, \alpha, k) + o\left(\frac{1}{r}\right), \quad r \to \infty, \ \beta := \frac{x}{r}. \qquad (11)$$

The equations for finding σ_1 are:

$$\int_{S_1} g(s,t)\sigma_1(t)dt = -u_0(s), \qquad (12)$$

$$u_{0N} - \zeta_1 u_0 + \frac{A\sigma_1 - \sigma_1}{2} - \zeta_1 \int_{S_1} g(s,t)\sigma_1(t)dt = 0, \qquad (13)$$

$$u_{0N} + \frac{A\sigma_1 - \sigma_1}{2} = 0, \qquad (14)$$

respectively, for conditions (6)–(8). The operator A is defined as follows:

$$A\sigma := 2 \int_{S_1} \frac{\partial}{\partial N_s} g(s,t)\sigma_1(t)dt. \qquad (15)$$

Equations (12)–(14) are uniquely solvable, but there are no analytic formulas for their solutions for bodies of arbitrary shapes. However, if the body D_1 is small, $ka \ll 1$, one can rewrite (10) as

$$u(x) = u_0(x) + g(x,0)Q_1 + \int_{S_1} [g(x,t) - g(x,0)]\sigma_1(t)dt, \qquad (16)$$

where

$$Q_1 := \int_{S_1} \sigma_1(t)dt, \qquad (17)$$

and $0 \in D_1$ is the origin.

If $ka \ll 1$, then we prove that

$$|g(x,0)Q_1| \gg \left| \int_{S_1} [g(x,t) - g(x,0)]\sigma_1(t)dt \right|, \quad |x| > a. \qquad (18)$$

Therefore, *the scattered field is determined outside D_1 by a single number Q_1.*

This number can be obtained analytically without solving equations (12)–(13). The case (14) requires a special approach by the reason discussed in detail later.

Let us give the results for equations (12) and (13) first. For equation (12) one has

$$Q_1 = \int_{S_1} \sigma_1(t)dt = -Cu_0(0)[1 + o(1)], \quad a \to 0, \tag{19}$$

where C *is the electric capacitance of a perfect conductor with the shape* D_1. For equation (13) one has

$$Q_1 = -\zeta|S_1|u_0(0)[1 + o(1)], \quad a \to 0, \tag{20}$$

where $|S_1|$ is the surface area of S_1. The scattering amplitude for problems (1)–(3) with $\Gamma = \Gamma_1$ (acoustically soft particle) is

$$A_1(\beta, \alpha) = -\frac{C}{4\pi}[1 + o(1)], \tag{21}$$

since

$$u_0(0) = e^{ik\alpha \cdot x}|_{x=0} = 1.$$

Therefore, in this case the scattering is isotropic and of the order $O(a)$, *because the capacitance* $C = O(a)$.

The scattering amplitude for problems (1)–(3) with $\Gamma = \Gamma_2$ (*small impedance particles*) is:

$$A_2(\beta, \alpha) = -\frac{\zeta_1|S_1|}{4\pi}[1 + o(1)], \tag{22}$$

since $u_0(0) = 1$.

In this case the scattering is also isotropic, and of the order $O(\zeta|S_1|)$.

If $\zeta_1 = O(1)$, then $A_2 = O(a^2)$, because $|S_1| = O(a^2)$. If $\zeta_1 = O\left(\frac{1}{a^\kappa}\right)$, $\kappa \in (0, 1)$, then $A_2 = O(a^{2-\kappa})$. The case $\kappa = 1$ was considered in [10].

The scattering amplitude for problems (1)–(3) with $\Gamma = \Gamma_3$ (acoustically hard particles) is

$$A_3(\beta, \alpha) = -\frac{k^2|D_1|}{4\pi}(1 + \beta_{pq}\beta_p\alpha_q), \quad \text{if } u_0 = e^{ik\alpha \cdot x}. \tag{23}$$

Here and below summation is understood over the repeated indices, $\alpha_q = \alpha \cdot e_q$, $\alpha \cdot e_q$ denotes the dot product of two vectors in $\mathbb{R}^3$, $p, q = 1, 2, 3$, $\{e_p\}$ is an orthonormal Cartesian basis of $\mathbb{R}^3$, $|D_1|$ is the volume of D_1, β_{pq} is the magnetic polarizability tensor defined as follows ([8], p. 62):

$$\beta_{pq} := \frac{1}{|D_1|} \int_{S_1} t_p \sigma_{1q}(t) dt, \tag{24}$$

σ_{1q} is the solution to the equation

$$\sigma_{1q}(s) = A_0 \sigma_{1q} - 2N_q(s), \tag{25}$$

$N_q(s) = N(s) \cdot e_q$, $N = N(s)$ is the unit outer normal to S_1 at the point s, i.e., the normal pointing out of D_1, and A_0 is the operator A at $k = 0$. For small bodies $\|A - A_0\| = o(ka)$.

If $u_0(x)$ is an arbitrary field satisfying equation (1), not necessarily the plane wave $e^{ik\alpha \cdot x}$, then

$$A_3(\beta, \alpha) = \frac{|D_1|}{4\pi} \left(ik\beta_{pq} \frac{\partial u_0}{\partial x_q} \beta_p + \triangle u_0 \right). \tag{26}$$

The above formulas are derived in Section 2. In Section 3, we develop a theory for many-body wave scattering problem and derive the equations for effective field in the medium, in which many small particles are embedded, as $a \to 0$.

The results, presented in this paper, are based on the earlier works of the author ([7]–[32]). These results and methods of their derivation differ much from those published by other authors.

Our approach to homogenization-type theory is also different from the approaches of other authors (see, for example, [4], [6]). The differences are:

(i) no periodic structure in the problems is assumed,
(ii) the operators in our problems are non-selfadjoint and have continuous spectrum,
(iii) the limiting medium is not homogeneous and its parameters are not periodic,
(iv) the technique for passing to the limit is different from one used in homogenization theory.

Let us summarize the results for one-body wave scattering.

Theorem 1.1. *The scattering amplitude for the problems (1)–(4) for small body of an arbitrary shape are given by formulas (25), (26), (27), for the boundary conditions Γ_1, Γ_2, Γ_3, respectively.*

2 Derivation of the formulas for one-body wave scattering problems

Let us recall the known result (see e.g., [7])

$$\frac{\partial}{\partial N_s^-} \int_{S_1} g(x,t)\sigma_1(t)dt = \frac{A\sigma_1 - \sigma_1}{2} \tag{27}$$

concerning the limiting value of the normal derivative of single-layer potential from outside. Let $x_m \in D_m$, $t \in S_m$, S_m is the surface of D_m, $a = 0.5 \operatorname{diam} D_m$.

In this section, $m = 1$, and $x_m = 0$ is the origin.

We assume that $ka \ll 1$, $ad^{-1} \ll 1$, so $|x - x_m| = d \gg a$. Then

$$\frac{e^{ik|x-t|}}{4\pi|x-t|} = \frac{e^{ik|x-x_m|}}{4\pi|x-x_m|}e^{-ik(x-x_m)^o\cdot(t-x_m)}\left(1 + O\left(ka + \frac{a}{d}\right)\right), \tag{28}$$

$$k|x-t| = k|x-x_m| - k(x-x_m)^o\cdot(t-x_m) + O\left(\frac{ka^2}{d}\right), \tag{29}$$

where

$$d = |x - x_m|, \quad (x - x_m)^o := \frac{x - x_m}{|x - x_m|},$$

and

$$\frac{|x-t|}{|x-x_m|} = 1 + O\left(\frac{a}{d}\right). \tag{30}$$

Let us derive estimate (19). Since $|t| \le a$ on S_1, one has

$$g(s,t) = g_0(s,t)(1 + O(ka)),$$

where $g_0(s,t) = \frac{1}{4\pi|s-t|}$. Since $u_0(s)$ is a smooth function, one has $|u_0(s) - u_0(0)| = O(a)$. Consequently, equation (12) can be considered as an equation for electrostatic charge distribution $\sigma_1(t)$ on the surface S_1 of a perfect conductor D_1, charged to the constant potential $-u_0(0)$ (up to a small term of the order $O(ka)$). It is known that the total charge $Q_1 = \int_{S_1} \sigma_1(t)dt$ of this conductor is equal to

$$Q_1 = -Cu_0(0)(1 + O(ka)), \tag{31}$$

where C is the electric capacitance of the perfect conductor with the shape D_1.

Analytic formulas for electric capacitance C of a perfect conductor of an arbitrary shape, which allow to calculate C with a desired accuracy, are derived in [8]. For example, the zeroth approximation formula is:

$$C^{(0)} = \frac{4\pi|S_1|^2}{\int_{S_1}\int_{S_1}\frac{dsdt}{r_{st}}}, \quad r_{st} = |t - s|, \tag{32}$$

and we assume in (32) that $\epsilon_0 = 1$, where ϵ_0 is the dielectric constant of the homogeneous medium in which the perfect conductor is placed. Formula (31) is formula (19). If $u_0(x) = e^{ik\alpha \cdot x}$, then $u_0(0) = 1$, and $Q_1 = -C(1 + O(ka))$. In this case

$$A_1(\beta, \alpha) = \frac{Q_1}{4\pi} = -\frac{C}{4\pi}[1 + O(ka)],$$

which is formula (21).

Consider now wave scattering by an impedance particle.

Let us derive formula (20). Integrate equation (13) over S_1, use the divergence formula

$$\int_{S_1} u_{0N}ds = \int_{D_1} \nabla^2 u_0 dx = -k^2 \int_{D_1} u_0 dx = k^2|D_1|u_0(0)[1 + o(1)], \tag{33}$$

where $|D_1| = O(a^3)$, and the formula

$$-\zeta_1 \int_{S_1} u_0 ds = -\zeta_1|S_1|u_0(0)[1 + o(1)], \tag{34}$$

which is valid because the body D_1 is small: in this case $u_0(s) \approx u_0(0)$. Furthermore $|\int_{S_1} g(s,t)ds| = O(a)$, so

$$\zeta_1 \int_{S_1} ds \int_{S_1} g(s,t)\sigma_1(t)dt = O(aQ_1). \tag{35}$$

Therefore, the term (35) is negligible compared with Q_1 as $a \to 0$. Finally, if $ka \ll 1$, then $g(s, t) = g_0(s, t)\left(1 + ik|s - t| + \cdots\right)$, and

$$\frac{\partial}{\partial N_s} g(s, t) = \frac{\partial}{\partial N_s} g_0(s, t)[1 + O(ka)]. \tag{36}$$

Denote by A_0 the operator

$$A_0 \sigma = 2 \int_{S_1} \frac{\partial g_0(s, t)}{\partial N_s} \sigma_1(t) dt. \tag{37}$$

It is known from the potential theory (see, for example, [31]) that

$$\int_{S_1} A_0 \sigma_1 ds = - \int_{S_1} \sigma_1(t) dt, \quad 2 \int_{S_1} \frac{\partial g_0(s, t)}{\partial N_s} ds = -1, \quad t \in S_1. \tag{38}$$

Therefore,

$$\int_{S_1} ds \frac{A\sigma_1 - \sigma_1}{2} = -Q_1[1 + O(ka)]. \tag{39}$$

Consequently, from formulas (33)–(39) one gets formula (22).

One can see that *the wave scattering by an impedance particle is isotropic, and the scattered field is of the order $O(\zeta_1|S_1|)$. Since $|S_1| = O(a^2)$, one has $O(\zeta_1|S_1|) = O(a^{2-\kappa})$ if $\zeta_1 = O\left(\frac{1}{a^\kappa}\right)$, $\kappa \in [0, 1)$.*

Consider now wave scattering by an acoustically hard small particle, i.e., the problem with the Neumann boundary condition.

In this case we will prove that:

(i) The scattering is anisotropic,
(ii) It is defined not by a single number, as in the previous two cases, but by a tensor,
    and
(iii) The order of the scattered field is $O(a^3)$ as $a \to 0$, for a fixed $k > 0$, i.e., the scattered field is much smaller than in the previous two cases.

Integrating over S_1 equation (14), one gets

$$Q_1 = \int_{D_1} \nabla^2 u_0 dx = \nabla^2 u_0(0)|D_1|[1 + o(1)], \quad a \to 0. \tag{40}$$

Thus, $Q_1 = O(a^3)$. Therefore, the contribution of the term $e^{-ikx^o \cdot t}$ in formula (28) with $x_m = 0$ will be also of the order $O(a^3)$ and should be taken into account, *in contrast to the previous two cases.* Namely,

$$u(x) = u_0(x) + g(x, 0) \int_{S_1} e^{-ik\beta \cdot t} \sigma_1(t) dt, \quad \beta := \frac{x}{|x|} = x^o. \quad (41)$$

One has

$$\int_{S_1} e^{-ik\beta \cdot t} \sigma_1(t) dt = Q_1 - ik\beta_p \int_{S_1} t_p \sigma_1(t) dt, \quad (42)$$

where the terms of higher order of smallness are neglected and summation over index p is understood. The function σ_1 solves equation (14):

$$\sigma_1 = A\sigma_1 + 2u_{0N} = A\sigma_1 + 2ik\alpha_q N_q u_0(s), \quad s \in S_1 \quad (43)$$

if $u_0(x) = e^{ik\alpha \cdot x}$.

Comparing (43) with (25), using (24), and taking into account that $ka \ll 1$, one gets

$$-ik\beta_p \int_{S_1} t_p \sigma_1(t) dt = -ik\beta_p |D_1| \beta_{pq} (-ik\alpha_q) u_0(0)[1 + O(ka)]$$

$$= -k^2 |D_1| \beta_{pq} \beta_p \alpha_q u_0(0)[1 + O(ka)]. \quad (44)$$

From (40), (42) and (44) one gets formula (23), because $\nabla^2 u_0 = -k^2 u_0$.

If $u_0(x)$ is an arbitrary function, satisfying equation (1), then $ik\alpha_q$ in (43) is replaced by $\frac{\partial u_0}{\partial x_q}$, and $-k^2 u_0 = \Delta u_0$, which yields formula (26).

This completes the derivation of the formulas for the solution of scalar wave scattering problem by one small body on the boundary of which the Dirichlet, or the impedance, or the Neumann boundary condition is imposed.

3 Many-body scattering problem

In this section, we assume that there are $M = M(a)$ small bodies (particles) D_m, $1 \le m \le M$, $a = 0.5 \max \operatorname{diam} D_m$, $ka \ll 1$. The distance $d = d(a)$ between neighboring bodies is much larger than

a, $d \gg a$, but *we do not assume that* $d \gg \lambda$, *so there may be many small particles on the distances of the order of the wavelength* λ.

This means that our medium with the embedded particles is not necessarily diluted.

We assume that the small bodies are embedded in an arbitrary large but finite domain D, $D \subset \mathbb{R}^3$, so $D_m \subset D$. Denote $D' := \mathbb{R}^3 \setminus D$ and $\Omega := \cup_{m=1}^{M} D_m$, $S_m := \partial D_m$, $\partial \Omega = \cup_{m=1}^{M} S_m$. By N we denote a unit normal to $\partial \Omega$, pointing out of Ω, by $|D_m|$ the volume of the body D_m is denoted.

The scattering problem consists of finding the solution to the following problem

$$(\nabla^2 + k^2)u = 0 \quad \text{in } \mathbb{R}^3 \setminus \Omega, \tag{45}$$

$$\Gamma u = 0 \quad \text{on } \partial \Omega, \tag{46}$$

$$u = u_0 + v, \tag{47}$$

where u_0 is the incident field, satisfying equation (45) in $\mathbb{R}^3$, for example, $u_0 = e^{ik\alpha \cdot x}$, $\alpha \in S^2$, and v is the scattered field, satisfying the radiation condition (5). The boundary condition (46) can be of the types (6)–(8).

In the case of impedance boundary condition (7) we assume that

$$u_N = \zeta_m u \quad \text{on } S_m, \quad 1 \le m \le M, \tag{48}$$

so the impedance may vary from one particle to another. We assume that

$$\zeta_m = \frac{h(x_m)}{a^\kappa}, \quad \kappa \in (0, 1), \tag{49}$$

where $x_m \in D_m$ is a point in D_m, and $h(x)$, $x \in D$, is a given function, which we can choose as we wish, subject to the condition $\text{Im} h(x) \le 0$. For simplicity we assume that $h(x)$ is a continuous function.

Let us make the following assumption about the distribution of small particles:

If $\Delta \subset D$ is an arbitrary open subset of D, then the number $\mathcal{N}(\Delta)$ of small particles in Δ, assuming the impedance boundary condition, is:

$$\mathcal{N}_\zeta(\Delta) = \frac{1}{a^{2-\kappa}} \int_\Delta N(x)dx[1 + o(1)], \quad a \to 0, \tag{50}$$

where $N(x) \geq 0$ is a given function.

If the Dirichlet boundary condition is assumed, then

$$\mathcal{N}_D(\Delta) = \frac{1}{a} \int_\Delta N(x)dx[1 + o(1)], \quad a \to 0. \tag{51}$$

The case of the Neumann boundary condition will not be considered in this chapter, see [30].

We look for the solution to problems (45)–(47) with the Dirichlet boundary condition of the form

$$u = u_0 + \sum_{m=1}^{M} \int_{S_m} g(x,t)\sigma_m(t)dt, \tag{52}$$

where $\sigma_m(t)$ are some functions to be determined from the boundary condition (46). It is proved in [10] that problems (45)–(47) have a unique solution of the form (52). For any $\sigma_m(t)$ function (52) solves equation (45) and satisfies condition (47). The boundary condition (46) determines σ_m uniquely. However, if $M \gg 1$, then numerical solution of the system of integral equations for σ_m, $1 \leq m \leq M$, which one gets from the boundary condition (46), is practically not feasible.

To avoid this principal difficulty, we prove that the solution to scattering problems (45)–(47) is determined by M numbers

$$Q_m := \int_{S_m} \sigma_m(t)dt, \tag{53}$$

rather than M functions $\sigma_m(t)$. This allows one to drastically reduce the complexity of the numerical solution of the many-body scattering problems in the case of small particles.

This is possible to prove if the particles D_m are small. We derive analytical formulas for Q_m as $a \to 0$.

Let us define the effective (self-consistent) field $u_e(x) = u_e^{(j)}(x)$, acting on the j-th particle, by the formula

$$u_e(x) := u(x) - \int_{S_j} g(x,t)\sigma_j(t)dt, \quad |x - x_j| \sim a. \tag{54}$$

Physically this field acts on the j-th particle and is a sum of the incident field and the fields acting from all other particles:

$$u_e(x) = u_e^{(j)}(x) := u_0(x) + \sum_{m \neq j} \int_{S_m} g(x,t)\sigma_m(t)dt. \tag{55}$$

Let us rewrite (55) as follows:

$$u_e(x) = u_0(x) + \sum_{m \neq j}^{M} g(x, x_m)Q_m + \sum_{m \neq j}^{M} \int_{S_m} [g(x,t) - g(x,x_m)]\sigma_m(t)dt. \tag{56}$$

We want to prove that the last sum is negligible compared with the first one as $a \to 0$.

To prove this, let us give some estimates. One has $|t - x_m| \leq a$, $d = |x - x_m|$,

$$|g(x,t) - g(x,x_m)| = \max\left\{O\left(\frac{a}{d^2}\right), O\left(\frac{ka}{d}\right)\right\},$$

$$|g(x,x_m)| = O(1/d). \tag{57}$$

Therefore, if $|x - x_j| = O(a)$, then

$$\frac{\left|\int_{S_m}[g(x,t) - g(x,x_m)]\sigma_m(t)dt\right|}{|g(x,x_m)Q_m|} \leq O(ad^{-1} + ka). \tag{58}$$

One can also prove that

$$J_1/J_2 = O(ka + ad^{-1}), \tag{59}$$

where J_1 is the first sum in (56) and J_2 is the second sum in (56). Therefore, at any point $x \in \Omega' = \mathbb{R}^3 \setminus \Omega$ one has

$$u_e(x) = u_0(x) + \sum_{m=1}^{M} g(x,x_m)Q_m, \quad x \in \Omega', \tag{60}$$

where the terms of higher order of smallness are omitted.

3.1   *The case of acoustically soft particles*

If (46) is the Dirichlet condition, then, as we have proved in Section 2
(see formula (31)), one has

$$Q_m = -C_m u_e(x_m). \tag{61}$$

Thus,

$$u_e(x) = u_0(x) - \sum_{m=1}^{M} g(x, x_m) C_m u_e(x_m), \quad x \in \Omega'. \tag{62}$$

One has

$$u(x) = u_e(x) + o(1), \quad a \to 0, \tag{63}$$

so the full field and effective field are practically the same.

Let us write a linear algebraic system (LAS) for finding unknown
quantities $u_e(x_m)$:

$$u_e(x_j) = u_0(x_j) - \sum_{m \neq j}^{M} g(x_j, x_m) C_m u_e(x_m). \tag{64}$$

If M is not very large, say $M = O(10^3)$, then LAS (64) can be solved
numerically, and formula (62) can be used for calculation of $u_e(x)$.

Consider the limiting case, when $a \to 0$. One can rewrite (64) as
follows:

$$u_e(\xi_q) = u_0(\xi_q) - \sum_{p \neq q}^{P} g(\xi_q, \xi_p) u_e(\xi_p) \sum_{x_m \in \Delta_p} C_m, \tag{65}$$

where $\{\Delta_p\}_{p=1}^{P}$ is a union of cubes which forms a covering of D,

$$\max_p diam \Delta_p := b = b(a) \gg a,$$

$$\lim_{a \to 0} b(a) = 0. \tag{66}$$

By $|\Delta_p|$ we denote the volume (measure) of Δ_p, and ξ_p is the center of Δ_p, or a point x_p in an arbitrary small body D_p, located in Δ_p. Let us assume that there exists the limit

$$\lim_{a \to 0} \frac{\sum_{x_m \in \Delta_p} C_m}{|\Delta_p|} = C(\xi_p), \quad \xi_p \in \Delta_p. \tag{67}$$

For example, one may have

$$C_m = c(\xi_p)a \tag{68}$$

for all m such that $x_m \in \Delta_p$, where $c(x)$ is some function in D. If all D_m are balls of radius a, then $c(x) = 4\pi$. We have

$$\sum_{x_m \in \Delta_p} C_m = C_p a \mathcal{N}(\Delta_p) = C_p N(\xi_p)|\Delta_p|[1 + o(1)], \quad a \to 0, \tag{69}$$

so limit (67) exists, and

$$C(\xi_p) = c(\xi_p)N(\xi_p). \tag{70}$$

From (65), (68)–(70) one gets

$$u_e(\xi_q) = u_0(\xi_q) - \sum_{p \neq q} g(\xi_q, \xi_p)c(\xi_p)N(\xi_p)u_e(\xi_p)|\Delta_p|, \quad 1 \leq p \leq P. \tag{71}$$

Linear algebraic system (71) can be considered as the *collocation method for solving integral equation*

$$u(x) = u_0(x) - \int_D g(x, y)c(y)N(y)u(y)dy. \tag{72}$$

It is proved in [25] that

System (71) is uniquely solvable for all sufficiently small $b(a)$, and the function

$$u_P(x) := \sum_{p=1}^{P} \chi_p(x)u_e(\xi_p) \tag{73}$$

converges in $L^\infty(D)$ to the unique solution of equation (72).

The function $\chi_p(x)$ in (73) is the characteristic function of the cube Δ_p: it is equal to 1 in Δ_p and vanishes outside Δ_p. Thus, if $a \to 0$, the solution to the many-body wave scattering problem in the case of the Dirichlet boundary condition is well approximated by the unique solution of the integral equation (72).

Applying the operator $L_0 := \nabla^2 + k^2$ to (72), and using the formula $L_0 g(x, y) = -\delta(x - y)$, where $\delta(x)$ is the delta-function, one gets

$$\nabla^2 u + k^2 u - q(x)u = 0 \quad \text{in } \mathbb{R}^3, \quad q(x) := c(x)N(x). \tag{74}$$

The physical conclusion is:

If one embeds $M(a) = O(1/a)$ small acoustically soft particles, which are distributed as in (51), then one creates, as $a \to 0$, a limiting medium, which is inhomogeneous, and has a refraction coefficient $n^2(x) = 1 - k^{-2}q(x)$.

It is interesting from the physical point of view to note that

The limit, as $a \to 0$, of the total volume of the embedded particles is zero.

Indeed, the volume of one particle is $O(a^3)$, the total number M of the embedded particles is $O(a^3 M) = O(a^2)$, and $\lim_{a \to 0} O(a^2) = 0$.

The second observation is: if (51) holds, then on a unit length straight line there are $O(\frac{1}{a^{1/3}})$ particles, so the distance between neighboring particles is $d = O(a^{1/3})$. If $d = O(a^\gamma)$ with $\gamma > \frac{1}{3}$, then the number of the embedded particles in a subdomain Δ_p is $O(\frac{1}{d^3}) = O(a^{-3\gamma})$. In this case, for $3\gamma > 1$, the limit in (69) is $C(\xi_p) = \lim_{a \to 0} c_p a O(a^{-3\gamma}) = \infty$. Therefore, the product of this limit by u remains finite only if $u = 0$ in D. Physically this means that if the distances between neighboring perfectly soft particles are smaller than $O(a^{1/3})$, namely, they are $O(a^\gamma)$ with any $\gamma > \frac{1}{3}$, then $u = 0$ in D.

On the other hand, if $\gamma < \frac{1}{3}$, then the limit $C(\xi_p) = 0$, and $u = u_0$ in D, so that the embedded particles do not change, in the limit $a \to 0$, properties of the medium.

This concludes our discussion of the scattering problem for many acoustically soft particles.

3.2 Wave scattering by many impedance particles

We assume now that (49) and (50) hold, use the exact boundary condition (46) with $\Gamma = \Gamma_2$, that is,

$$u_{eN} - \zeta_m u_e + \frac{A_m \sigma_m - \sigma_m}{2} - \zeta_m \int_{S_m} g(s,t)\sigma_m(t)dt = 0, \qquad (75)$$

and integrate (75) over S_m in order to derive an analytical asymptotic formula for $Q_m = \int_{S_m} \sigma_m(t)dt$.

We have

$$\int_{S_m} u_{eN} ds = \int_{D_m} \nabla^2 u_e dx = O(a^3), \qquad (76)$$

$$\int_{S_m} \zeta_m u_e(s) ds = h(x_m)a^{-\kappa}|S_m|u_e(x_m)[1 + o(1)], \quad a \to 0, \qquad (77)$$

$$\int_{S_m} \frac{A_m \sigma_m - \sigma_m}{2} ds = -Q_m[1 + o(1)], \quad a \to 0, \qquad (78)$$

and

$$\zeta_m \int_{S_m} \int_{S_m} g(s,t)\sigma_m(t)dt = h(x_m)a^{1-\kappa}Q_m = o(Q_m), \quad 0 < \kappa < 1. \qquad (79)$$

From (75)–(79) one finds

$$Q_m = -h(x_m)a^{2-\kappa}|S_m|a^{-2}u_e(x_m)[1 + o(1)]. \qquad (80)$$

This yields the formula for the approximate solution to the wave scattering problem for many impedance particles:

$$u(x) = u_0(x) - a^{2-\kappa} \sum_{m=1}^{M} g(x, x_m)b_m h(x_m)u_e(x_m)[1 + o(1)], \qquad (81)$$

where

$$b_m := |S_m|a^{-2}$$

are some positive numbers which depend on the geometry of S_m and are independent of a. For example, if all D_m are balls of radius a, then $b_m = 4\pi$.

A linear algebraic system for $u_e(x_m)$, analogous to (64), is

$$u_e(x_j) = u_0(x_j) - a^{2-\kappa} \sum_{m=1, m \neq j}^{M} g(x_j, x_m) b_m h(x_m) u_e(x_m). \quad (82)$$

The integral equation for the limiting effective field in the medium with embedded small particles, as $a \to 0$, is

$$u(x) = u_0(x) - b \int_D g(x, y) N(y) h(y) u(y) dy, \quad (83)$$

where

$$u(x) = \lim_{a \to 0} u_e(x), \quad (84)$$

and we have assumed in (83) for simplicity that $b_m = b$ for all m, that is, all small particles are of the same shape and size.

Applying operator $L_0 = \nabla^2 + k^2$ to equation (83), one finds the differential equation for the limiting effective field $u(x)$:

$$(\nabla^2 + k^2 - bN(x)h(x))u = 0 \quad \text{in } \mathbb{R}^3, \quad (85)$$

and u satisfies condition (47).

The conclusion is: the limiting medium is inhomogeneous, and its properties are described by the function

$$q(x) := bN(x)h(x). \quad (86)$$

This concludes our discussion of the wave scattering problem with many small impedance particles.

4   Creating materials with a desired refraction coefficient

Since the choice of the functions $N(x) \geq 0$ and $h(x)$, $\text{Im}h(x) \leq 0$, is at our disposal, we can create the medium with a desired refraction coefficient by embedding many small impedance particles, with suitable impedances, according to the distribution law (50) with a suitable $N(x)$. The function

$$n_0^2(x) - k^{-2}q(x) = n^2(x) \quad (87)$$

is the refraction coefficient of the limiting medium, where $n_0^2(x)$ is the refraction coefficient of the original medium (see also Section 5). In equation (85) it is assumed that $n_0^2(x) = 1$. If $n_0^2(x) \neq 1$, then the operator $L_0 = \nabla^2 + k^2 n_0^2(x)$.

A recipe for creating material with a desired refraction coefficient can now be formulated.

Given a desired refraction coefficient $n^2(x)$, $\text{Im}n^2(x) \geq 0$, one can find $N(x)$ and $h(x)$ so that (87) holds, where $q(x)$ is defined in (86), that is, one can create a material with a desired refraction coefficient by embedding into a given material many small particles with suitable boundary impedances and suitable distribution law.

5   Scattering by small particles embedded in an inhomogeneous medium

Suppose that the operator $\nabla^2 + k^2$ in (1) and in (45) is replaced by the operator $L_0 = \nabla^2 + k^2 n_0^2(x)$, where $n_0^2(x)$ is a known function,

$$\text{Im}n_0^2(x) \geq 0. \tag{88}$$

The function $n_0^2(x)$ is the refraction coefficient of an inhomogeneous medium in which many small particles are embedded. The results, presented in Sections 1–3 remain valid if one replaces function $g(x, y)$ by the Green's function $G(x, y)$,

$$[\nabla^2 + k^2 n_0^2(x)]G(x, y) = -\delta(x - y), \tag{89}$$

satisfying the radiation condition. We assume that

$$n_0^2(x) = 1 \quad \text{in } D' := \mathbb{R}^3 \setminus D. \tag{90}$$

The function $G(x, y)$ is uniquely defined (see, e.g., [10]). The derivations of the results remain essentially the same because

$$G(x, y) = g_0(x, y)[1 + O(|x - y|)], \quad |x - y| \to 0, \tag{91}$$

where $g_0(x, y) = \frac{1}{4\pi|x-y|}$. Estimates of $G(x, y)$ as $|x - y| \to 0$ and as $|x - y| \to \infty$ are obtained in [10]. Smallness of particles in an inhomogeneous medium with refraction coefficient $n_0^2(x)$ is described by the relation $k n_0 a \ll 1$, where $n_0 := \max_{x \in D} |n_0(x)|$, and $a = \max_{1 \leq m \leq M} \operatorname{diam} D_m$.

6   Conclusions

Analytic formulas for the scattering amplitudes for wave scattering by a single small particle are derived for small acoustically soft, or hard, or impedance particles.

The equation for the effective field in the medium, in which many small particles are embedded, is derived in the limit $a \to 0$. *The physical assumptions $a \ll d \ll \lambda$ are such that the multiple scattering effects are essential.* The derivations are rigorous.

On the basis of the developed theory efficient numerical methods are proposed for solving many-body wave scattering problems in the case of small scatterers. These methods allow one to solve the problems which earlier were not possible to solve.

A method for creating materials with a desired refraction coefficient is given and rigorously justified. Its practical implementation requires development of a method for preparing small particles with prescribed boundary impedances.

The physically novel point, compared with the known results for wave scattering by small bodies, is the dependence on the size a of the small scatterer which is much larger than $O(a^3)$, the Rayleigh-type dependence, see, for example, formula (22), where the dependence on a is $O(\zeta|S_1|) = O(a^{2-\kappa})$. The formulas for the wave scattering by small particles of an arbitrary shape for various types of the boundary conditions are new. The equations for the effective field in the medium, in which many small particles with various boundary conditions are embedded, are new.

In this paper we did not discuss the EM (electromagnetic waves) scattering and the related problems of creating materials with a desired refraction coefficient. See [31], [33], [34].

References

[1]   Andriychuk, M., Ramm, A.G., Numerical solution of many-body wave scattering problem for small particles and creating materials with desired refraction coefficient, Chapter in the book: "Numerical

Simulations of Physical and Engineering Processes", InTech., Vienna, 2011, pp. 1–28. (edited by Jan Awrejcewicz) ISBN 978-953-307-620-1

[2] Andriychuk, M., Ramm, A.G., Scattering of electromagnetic waves by many thin cylinders: theory and computational modeling, Optics Communications, 285, N20, (2012), 4019–4026.

[3] H.C. Hulst van de, Light scattering by small particles, Dover, New York, 1961.

[4] V. Jikov, S. Kozlov, O. Oleinik, Homogenization of differential operators and integral functionals, Springer-verlag, Berlin, 1994.

[5] L. Landau, L. Lifschitz, Electrodynamics of continuous media, Pergamon Press, Oxford, 1984.

[6] V. Marchenko, E. Khruslov, Homogenization of partial differential equations, Birkhäuser, Boston, 2006.

[7] A.G. Ramm, Scattering by obstacles, D.Reidel, Dordrecht, 1986.

[8] A.G. Ramm, Wave scattering by small bodies of arbitrary shapes, World Sci. Publishers, Singapore, 2005.

[9] A.G. Ramm, Scattering by many small bodies and applications to condensed matter physics, Europ. Phys. Lett., 80, 44001, (2007).

[10] A.G. Ramm, Many-body wave scattering by small bodies and applications, J. Math. Phys., 48, 103511, (2007).

[11] A.G. Ramm, Wave scattering by small particles in a medium, Phys. Lett. A 367, 156–161, (2007).

[12] A.G. Ramm, Wave scattering by small impedance particles in a medium, Phys. Lett. A 368, 164–172, (2007).

[13] A.G. Ramm, Distribution of particles which produces a "smart" material, Jour. Stat. Phys., 127, 915–934, (2007).

[14] A.G. Ramm, Distribution of particles which produces a desired radiation pattern, Physica B, 394, 253–255, (2007).

[15] A.G. Ramm, Creating wave-focusing materials, LAJSS (Latin-American Journ. of Solids and Structures), 5, 119–127, (2008).

[16] A.G. Ramm, Electromagnetic wave scattering by small bodies, Phys. Lett. A, 372, 4298–4306, (2008).

[17] A.G. Ramm, Wave scattering by many small particles embedded in a medium, Phys. Lett. A, 372, 3064–3070, (2008).

[18] A.G. Ramm, Preparing materials with a desired refraction coefficient and applications, In the book "Topics in Chaotic Systems: Selected Papers from Chaos 2008 International Conference", Editors C. Skiadas, I. Dimotikalis, Char. Skiadas, World Sci. Publishing, pp. 265–273, (2009).

[19] A.G. Ramm, Preparing materials with a desired refraction coefficient, Nonlinear Analysis: Theory, Methods and Appl., 70, e186-e190, (2009).

[20]　A.G. Ramm, Creating desired potentials by embedding small inhomogeneities, J. Math. Phys., 50, 123525, (2009).

[21]　A.G. Ramm, A method for creating materials with a desired refraction coefficient, Internat. Journ. Mod. Phys. B, 24, 5261–5268, (2010).

[22]　A.G. Ramm, Materials with a desired refraction coefficient can be created by embedding small particles into the given material, International Journal of Structural Changes in Solids (IJSCS), 2, 17–23, (2010).

[23]　A.G. Ramm, Wave scattering by many small bodies and creating materials with a desired refraction coefficient, Afrika Matematika, 22, 33–55, (2011).

[24]　A.G. Ramm, Scattering by many small inhomogeneities and applications, In the book "Topics in Chaotic Systems: Selected Papers from Chaos 2010 International Conference", Editors C. Skiadas, I. Dimotikalis, Char. Skiadas, World Sci. Publishing, pp. 41–52, (2011),

[25]　A.G. Ramm, Collocation method for solving some integral equations of estimation theory, Internat. Journ. of Pure and Appl. Math., 62, 57–65, (2010).

[26]　A.G. Ramm, Scattering of scalar waves by many small particles, AIP Advances, 1, 022135, (2011).

[27]　A.G. Ramm, Scattering of electromagnetic waves by many thin cylinders, Results in Physics, 1, N1, (2011), 13–16.

[28]　A.G. Ramm, Electromagnetic wave scattering by many small perfectly conducting particles of an arbitrary shape, Optics Communications, 285, N18, (2012), 3679–3683.

[29]　A.G. Ramm, Wave scattering by many small bodies: transmission boundary conditions, Reports on Math. Physics, 71, N3, (2013), 279–290.

[30]　A.G. Ramm, Many-body wave scattering problems in the case of small scatterers, Journal of Appl. Math and Comput., (JAMC), 41, N1, (2013), 473–500.

[31]　A.G. Ramm, Scattering of Acoustic and Electromagnetic Waves by Small Bodies of Arbitrary Shapes. Applications to Creating New Engineered Materials, Momentum Press, New York, 2013.

[32]　A.G. Ramm, N. Tran, A fast algorithm for solving scalar wave scattering problem by billions of particles, Journal of Algorithms and Optimization, 3, N1, (2015), 1–13.

[33] A.G. Ramm, Scattering of electromagnetic waves by many nano-wires, Mathematics, 1, (2013), 89–99.

[34] A.G. Ramm, Scattering of EM waves by many small perfectly conducting or impedance bodies, J. Math. Phys. (JMP), 56, N9, 091901, (2015).

[35] J. Rayleigh, Scientific papers, Cambridge Univ. Press, Cambridge, 1992.

Appendix C: Wave Scattering by Many Small Impedance Particles and Applications*

Alexander G. Ramm

Department of Mathematics
Kansas State University, Manhattan, KS 66506-2602, USA
ramm@ksu.edu

Abstract

Formulas are derived for solutions of many-body wave scattering problem by small impedance particles embedded in a homogeneous medium. The limiting case is considered, when the size a of small particles tends to zero while their number tends to infinity at a suitable rate. The basic physical assumption is $a \ll d \ll \lambda$, where d is the minimal distance between neighboring particles, λ is the wavelength, and the particles can be impedance balls $B(x_m, a$ with centers x_m located on a grid. Equations for the limiting effective (self-consistent) field in the medium are derived. It is proved that one can creat material with a desired refraction coefficient by embedding in a free space many small balls of radius a with prescribed boundary impedances. The small balls can be centered at the points located on a grid. A recipe for creating materials with a desired refraction coefficient is formulated. It is proved that materials with a desired radiation pattern, for example, wave-focusing materials, can be created.

---

*This chapter is published in *Reports on Mathematical Physics*, **90**(2), 193–202 (2022).

Keywords: wave scattering by many small bodies; smart materials.

1 Introduction

There is a large literature on wave scattering by small bodies, starting from Rayleigh's work (1871), [36], [1], [2]. For the problem of wave scattering by one body, an analytical solution was found only for the bodies of special shapes, for example, for balls and ellipsoids. If the scatterer is small, then the scattered field can be calculated analytically for bodies of arbitrary shapes, see [5], where this theory is presented.

The many-body wave scattering problem was discussed in the literature mostly numerically, if the number of scatterers is small, or under the assumption that the influence of the waves, scattered by other particles on a particular particle is negligible (see [3], where one finds a large bibliography, 1386 entries). This corresponds to the case when *the distance d between neighboring particles is much larger than the wavelength λ, and the characteristic size a of a small body (particle) is much smaller than λ*. Theoretically and practically the assumptions $a \ll \lambda$, $d \gg \lambda$ are the simplest and they allow to neglect multiple scattering. By $k = \frac{2\pi}{\lambda}$ the wave number is denoted.

In contrast, in our theory the basic assumption is $a \ll d \ll \lambda$, and *the multiple scattering is of basic importance*. We give references to our papers and monographs in which the theory of wave scattering by small bodies of arbitrary shapes was developed under the assumption $a \ll d \ll \lambda$, [4]–[34]. The novelty of the results in this paper is in the location of the small bodies: *they are placed on a grid*. This may be of practical interest. In reference [35], for the first time, the scattering problem for 10 billions small particles is solved numerically and numerical results are presented.

This paper is a presentation of the new results under simplifying assumptions: the small particles $D_m = B(x_m, a)$, $1 \leq m \leq M$, are impedance balls with prescribed boundary impedances ζ_m; the centers x_m of the balls are placed on a grid and are embedded in a homogeneous space in a bounded domain D, for example, in a box.

The basic results of this paper consist of:

(i) Solution to *many-body wave scattering problem* by small impedance particles, embedded in a homogeneous medium, under the assumptions $a \ll d \ll \lambda$, where d is the minimal distance between neighboring particles and λ is the wavelength in this medium;

(ii) Derivation of the equations for the limiting effective (self-consistent) field in this medium, in which many small impedance particles are embedded, when $a \to 0$ and the number $M = M(a)$ of the small particles tends to infinity at an appropriate rate;

(iii) Derivation of linear algebraic systems (LAS) for solving many-body wave scattering problems. These systems are not obtained by a discretization of boundary integral equations, and they give an efficient numerical method for solving many-body wave scattering problems in the case of small scatterers under the assumption $a \ll d \ll \lambda$;

(iv) Formulation of a recipe for creating materials with a desired refraction coefficient;

(v) Formulation of a method for creating materials with a desired radiation pattern.

Our methods give powerful numerical methods for solving many-body wave scattering problems in the case when the scatterers are small (see [31]).

Let us formulate the wave scattering problems we deal with. Let D be a bounded domain in $\mathbb{R}^3$ with a sufficiently smooth boundary. The scattering problem consists of finding the solution to the problem:

$$(\nabla^2 + k^2)u = 0 \quad \text{in } G' := \mathbb{R}^3 \setminus G, \quad G := \cup_{m=1}^M D_m, \quad k = \text{const} > 0,$$
$$(1)$$

where $D_m = B(x_m, a)$ is an impedance ball, centered at x_m and of small radius a,

$$u = u_0 + v, \quad u_0 = e^{ik\alpha \cdot x}, \quad \alpha \in S^2, \qquad (2)$$

S^2 is the unit sphere in $\mathbb{R}^3$, u_0 is the incident field, v is the scattered field satisfying the radiation condition

$$v_r - ikv = o\left(\frac{1}{r}\right), \quad r := |x| \to \infty, \quad v_r := \frac{\partial v}{\partial r}, \qquad (3)$$

and u satisfies the impedance boundary condition (bc) on the boundary of G:

$$u_N - \zeta_m u = 0, \quad \text{on } S_m, \quad \text{Im}\zeta_m \leq 0, \tag{4}$$

where ζ_m is a constant, N is the unit normal to $S := \cup_{m=1}^M S_m$, pointing out of $G := \cup_{m=1}^M D_m$, and S_m is the surface of $D_m = B(x_m, a)$.

By refraction coefficient $n(x)$ the coefficient in the equation

$$(\nabla^2 + k^2 n^2(x))u = (\nabla^2 + k^2 - q(x))u = 0 \tag{5}$$

is understood, where $q(x) := k^2(n^2(x) - 1)$.

Let $g(x, y) = \frac{e^{ik|x-y|}}{4\pi|x-y|}$. Then $(\nabla^2 + k^2)g(x, y) = -\delta(x - y)$, where $\delta(x)$ is the delta function.

Let us distribute small impedance particles $D_m = B(x_m, a)$ in D so that

$$\mathbb{N}(\Delta) = a^{\kappa-2}|\Delta|[1 + o(1)], \quad a \to 0, \tag{6}$$

where $\Delta \subset D$ is an arbitrary connected open subset of D, $|\Delta|$ is its volume, $\kappa \in (0, 1)$ is a number the experimenter may choose arbitrarily and $\mathbb{N}(\Delta)$ is the number of particles in Δ. Throughout this paper the important assumptions $a \ll d \ll \lambda$ and (6) are satisfied. As $a \to 0$ the number of small particles $\mathbb{N}(\Delta)$ in (6) tends to infinity since $\kappa - 2 < 0$.

The boundary impedances ζ_m are chosen by the formula

$$\zeta_m = a^{-\kappa}h(x_m), \tag{7}$$

where $h(x)$ is a continuous function in D, $\text{Im}h \leq 0$.

It will be clear from Section 3 that the function $h(x)$ can be determined by choosing a suitable boundary impedance $\zeta(x)$. When $a \to 0$, the ζ_m and $h(x_m)$ can be considered as continuous functions $\zeta(x)$ and $h(x)$.

The many-body scattering problems (1)–(4) have a solution and this solution is unique, see [31]. In Section 2, a method for solving this problem is given. In Section 3 a recipe for creating materials with a desired refraction coefficient is given. In Section 4, a recipe for creating materials with a desired radiation pattern is given.

2 Solution of many-body scattering problem

We look for the solution of the form

$$u = u_0 + \sum_{m=1}^{M} \int_{S_m} g(x,s)\sigma_m(s)ds = \sum_{m=1}^{M} g(x,x_m)Q_m + J, \quad (8)$$

where $\sigma_m(s)$ are unknown, $Q_m := \int_{S_m} \sigma_m(s)ds$. One may think about σ_m as of charge densities on S_m and of Q_m as of total charge on the surface S_m. We prove that $J := \sum_{m=1}^{M} \int_{S_m} [g(x,s) - g(x,x_m)]\sigma_m(s)ds$ is negligible compared to $I := \sum_{m=1}^{M} g(x,x_m)Q_m$, $J \ll I$ as $a \to 0$.

Let us prove this claim. First, we need the following lemma.

Lemma 1. *One has:*

$$Q_m = -4\pi a^2 \zeta_m u_m = -4\pi a^{2-\kappa} h_m u_m, \quad h_m := h(x_m), \quad u_m := u(x_m).$$
$$(9)$$

Proof. Let us define *the effective field acting on the m-th body:*

$$u_e := u_e^m := u - \int_{S_m} g(x,s)\sigma_m(s)ds.$$

If a is small, then $u(x) \sim u_e(x)$ for any x such that $|x - x_m| \geq d$. Let us use the exact boundary condition (4) for u_e and the known formula for the normal derivative of the single layer potential to get

$$u_{eN} + (A\sigma_m - \sigma_m)/2 - \zeta_m u_{em} - \zeta_m \int_{S_m} g(x,s)\sigma_m(s)ds = 0. \quad (10)$$

Here $A\sigma := \int_{S_m} g_{N_t}(t,s)\sigma_m(s)ds$, $t \in S_m$. Let us integrate (10) over S_m and keep the main term as $a \to 0$. One knows that $\int_{S_m} (A\sigma - \sigma)/2dt = -Q_m$. Furthermore, $\int_{S_m} g(t,s)ds = a$, as one can can check by a simple calculation using the fact that S_m is a sphere of radius a. This allows one to conclude that

$$\zeta_m \int_{S_m} ds\sigma_m(s) \int_{S_m} g(t,s)dt = h_m a^{1-\kappa} Q_m,$$

$$\zeta_m \int_{S_m} u_e ds = -4\pi a^{2-\kappa} h_m u_{em}$$

and $\int_{S_m} u_e N ds = O(a^2)$ as $a \to 0$. From the above estimates the conclusion of Lemma 1 follows. $\qquad\qquad\qquad\qquad\qquad\qquad\square$

Let us now check our claim $J \ll I$ as $a \to 0$. One has

$$g(x, x_m) Q_m = O(a^{2-\kappa} d^{-1})$$

for $|x - x_m| > d$, $a \to 0$. On the other hand, one derives:

$$\left| \int_{S_m} [g(t, s) - g(x, x_m)] \sigma_m(s) ds \right| \leq O(ad^{-2} a^{2-\kappa})$$

$$= O\left(\frac{a}{d}\right) O(a^{2-\kappa} d^{-1}).$$

This estimate justifies our claim since $a \ll d$. It follows that asymptotically, as $a \to 0$, one has

$$u \sim u_0 + \sum_{m=1}^{M} g(x, x_m) Q_m \sim u_0 - 4\pi a^{2-\kappa} \sum_{m=1}^{M} g(x, x_m) h_m u_m, \quad (11)$$

for $|x - x_m| \geq a$. Note that $M = O(a^{\kappa-2})$. Formula (11) allows one to calculate $u(x)$ at any point x, if the numbers u_m, $1 \leq m \leq M$, are known. One can use the following linear algebraic system (LAS) for finding u_m:

$$u_j = u_{0j} - 4\pi a^{2-\kappa} \sum_{m \neq j}^{M} g(x_j, x_m) h_m u_m, \quad 1 \leq j \leq M. \quad (12)$$

The order $M = O(a^{\kappa-2})$ of this system is large if a is small. One can reduce this order: consider a covering of D by non-intersecting small cubes Δ_p, $1 \leq p \leq P$, such that $d \ll diam(\Delta_p) \ll \lambda$, $u_m \sim u_p$, $h_m \sim h_p$ for all $x_m \in \Delta_p$. Then formula (12) can be written as

$$u_q = u_{0q} - 4\pi a^{2-\kappa} \sum_{p \neq q}^{P} g(x_q, x_p) h_p u_p \sum_{x_m \in \Delta_p} 1$$

$$= u_{0q} - 4\pi \sum_{p \neq q}^{P} g(x_q, x_p) h_p u_p |\Delta_p|, \quad (13)$$

where $a^{2-\kappa} \sum_{x_m \in \Delta_p} 1 = |\Delta_p|$ by formula (6). As $a \to 0$, $diam(\Delta_p) \to 0$ and formula (13) yields in the limit the integral equation for u:

$$u(x) = u_0(x) - 4\pi \int_D g(x, y)h(y)u(y)dy. \qquad (14)$$

Lemma 2. *Equation* (14) *has a solution, this solution is unique and it is a limiting value of the solution to the scattering problems* (1)–(4).

Proof. Apply the operator $\nabla^2 + k^2$ to equation (14) and get

$$(\nabla^2 + k^2)u = 4\pi h(x)u(x). \qquad (15)$$

This is a Schröedinger equation with potential $q(x) := 4\pi h(x)$; equations (2)–(3) hold. We assumed Im$h \leq 0$. Therefore, (15) has at most one solution. It is a Fredholm-type equation, so it has a solution. Lemma 2 is proved. $\qquad \square$

It follows from Lemma 2 that the LAS (13) for u_p is solvable and its solution is unique. Let us write equation (15) as

$$\nabla^2 u + k^2 n^2(x)u = 0, \quad n^2(x) := 1 - 4\pi k^{-2}h(x). \qquad (16)$$

Conclusion: *Embedding small impedance balls* $B(x_m, a)$ *in* D *results in creating in* D *a new material with the refraction coefficient*

$$n(x) = (1 - 4\pi k^{-2}h(x))^{1/2}. \qquad (17)$$

If one wants to have a material with the refraction coefficient $n(x)$, then one chooses by (17) the function $h(x)$. If $h(x)$ is chosen, then one knows the boundary impedance $\zeta(x)$ which generates the desired $h(x)$. The practical problem is to prepare small particles with the desired boundary impedance.

3   Recipe for creating materials with a desired refraction coefficient

Let us formulate a recipe for creating materials with a desired refraction coefficient. Formula (17) shows that if $h(x)$ is chosen properly, then any $n(x)$ can be obtained in D.

Recipe for creating materials with a desired refraction coefficient:

(a) *Calculate by formula (17) the function $h(x)$;*
(b) *Distribute small impedance balls in the domain D by the distribution law (6). The boundary impedances of these balls are defined by the function $h(x)$.*

Theorem 1. *The refraction coefficient of the resulting medium tends to the desired coefficient $n(x)$ as $a \to 0$.*

Let us show that a practically negative refraction coefficient $n(x)$ can be obtained by the above recipe. Denote $b := 4\pi k^{-2} > 0$ and write (17) as $n(x) = (1 - bh(x))^{1/2} = |1 - bh(x)|^{1/2} e^{\phi/2}$, where ϕ is the argument of $1 - bh(x)$. Since the operator in (14) is Fredholm, it remains Fredholm under small perturbations. Therefore one can take $h - i\epsilon$, where $\epsilon > 0$ is sufficiently small and equation (14) will still have a unique solution.

By choosing h so that $\text{Re}(1 - bh) > 0$ and $\text{Im}(1 - bh) < 0$ and small, one gets the argument $\phi = 2\pi - \delta$, where $\delta > 0$ is arbitrarily small if ϵ is sufficiently small. Then $n(x)$ will be nearly negative: its argument will be $\pi - \delta/2$.

4 Creating materials with a desired radiation pattern

Let is define what we mean by radiation pattern. Consider the scattering problem for equation (15):

$$\nabla^2 u + k^2 u - q(x)u = 0, \quad u = e^{ik\alpha \cdot x} + v, \qquad (18)$$

where v satisfies the radiation condition. Assume that $k > 0$ and $\alpha \in S^2$ are fixed. Then the scattering amplitude $A(\beta, \alpha, k) = A(\beta)$, where the dependence on k, α is dropped since k and α are fixed. The formula for the scattering amplitude is known, see, e.g., [34]:

$$A(\beta) := A_q(\beta) = -\frac{1}{4\pi} \int e^{-ik\beta \cdot y} q(y)u(y)dy. \qquad (19)$$

We call $A(\beta)$ the radiation pattern.

Consider an inverse problem (IP):

Given an arbitrary $f(\beta) \in L^2(S^2)$ and an arbitrary small $\epsilon > 0$, can one find a $q \in L^2(D)$ such that

$$\|f(\beta) - A_q(\beta)\|_{L^2(S^2)} < \epsilon. \tag{20}$$

Theorem 2. *For any $f(\beta) \in L^2(S^2)$ and an arbitrary small $\epsilon > 0$ there is a $q \in L^2(D)$ such that (20) holds.*

Since small perturbations of q result in small perturbations of $A(\beta)$, there are infinitely many potentials q for which inequality (20) holds.

The conclusion of Theorem 2 follows from Lemmas 3 and 4.

Lemma 3. *The set $\{\int_D e^{-ik\beta \cdot x} h(x) dx\}_{\forall h \in L^2(D)}$ is dense in $L^2(S^2)$.*

Corollary 1. *Given $f \in L^2(S^2)$ and $\epsilon > 0$, one can find $h \in L^2(D)$ such that*

$$\left\| f(\beta) + \frac{1}{4\pi} \int_D e^{-ik\beta \cdot x} h(x) dx \right\| < \epsilon.$$

Lemma 4. *The set $\{q(x)u(x, \alpha)\}_{\forall q \in L^2(D)}$ is dense in $L^2(D)$.*

Corollary 2. *Given $h \in L^2(D)$ and $\epsilon > 0$, one can find $q \in L^2(D)$ such that*

$$\|h(x) - q(x)u(x, \alpha)\|_{L^2(D)} < \epsilon.$$

Since the scattering amplitude

$$A(\beta) = -\frac{1}{4\pi} \int_D e^{-ik\beta \cdot x} h(x) dx$$

depends continuously on h, the inverse problem **IP** *is solved by Lemmas 3 and 4.*

Proof of Lemma 3. Assume the contrary. Then $\exists \psi \in L^2(S^2)$ such that

$$0 = \int_{S^2} d\beta \psi(\beta) \int_D e^{-ik\beta \cdot x} h(x) dx \quad \forall h \in L^2(D).$$

Thus,

$$\int_{S^2} d\beta \psi(\beta) e^{-ik\beta \cdot x} = 0 \quad \forall x \in \mathbb{R}^3.$$

Therefore,

$$\int_0^\infty d\lambda \lambda^2 \int_{S^2} d\beta e^{-i\lambda\beta \cdot x} \psi(\beta) \frac{\delta(\lambda - k)}{k^2} = 0 \quad \forall x \in \mathbb{R}^3.$$

By the injectivity of the Fourier transform, one gets

$$\psi(\beta) \frac{\delta(\lambda - k)}{k^2} = 0.$$

Therefore, $\psi(\beta) = 0$. Lemma 3 is proved.   □

Proof of Lemma 4. Given $h \in L^2(D)$, define

$$u := u_0 - \int_D g(x, y) h(y) dy, \quad g := \frac{e^{ik|x-y|}}{4\pi|x-y|}, \tag{21}$$

$$q(x) := \frac{h(x)}{u(x)}. \tag{22}$$

If $q \in L^2(D)$, then this q solves the problem, and u, defined in (21), is the scattering solution:

$$u = u_0 - \int_D g(x, y) q(y) u(y) dy, \tag{23}$$

and

$$A(\beta) = -\frac{1}{4\pi} \int_D e^{-ik\beta \cdot y} h(y) dy.$$

If q is not in $L^2(D)$, then the null set $N := \{x : x \in D, u(x) = 0\}$ is non-void. Let

$$N_\delta := \{x : |u(x)| < \delta, x \in D\}, \quad D_\delta := D \setminus N_\delta.$$

Claim 1.

$$\exists h_\delta = \begin{cases} h, & \text{in } D_\delta, \\ 0, & \text{in } N_\delta, \end{cases} \quad \text{such that } \|h_\delta - h\|_{L^2(D)} < c\epsilon,$$

$$q_\delta := \begin{cases} \dfrac{h_\delta}{u_\delta}, & \text{in } D_\delta, \\ 0, & \text{in } N_\delta, \end{cases} \quad q_\delta \in L^\infty(D), \quad u_\delta := u_0 - \int_D g h_\delta dy.$$

Proof of Claim 1. The set N is, generically, a line $l = \{x : u_1(x) = 0, \; u_2(x) = 0\}$, where $u_1 = \Re u$ and $u_2 = \Im u$. Consider a tubular neighborhood of this line, $\rho(x, l) \leq \delta$. Let the origin O be chosen on l, s_3 be the Cartesian coordinate along the tangent to l, and $s_1 = u_1$, $s_2 = u_2$ are coordinates in the plane orthogonal to l, s_j-axis is directed along $\nabla u_j|_l$, $j = 1, 2$.

The Jacobian $\mathcal{J}$ of the transformation $(x_1, x_2, x_3) \mapsto (s_1, s_2, s_3)$ is nonsingular, $|\mathcal{J}| + |\mathcal{J}^{-1}| \leq c$, because ∇u_1 and ∇u_2 are linearly independent. Define $h_\delta := \begin{cases} h, \text{ in } D_\delta, \\ 0, \text{ in } N_\delta, \end{cases}$ $u_\delta := u_0 - \int_D g(x, y) h_\delta(y) dy$,

$q_\delta := \begin{cases} \frac{h_\delta}{u_\delta}, \text{ in } D_\delta, \\ 0, \text{ in } N_\delta. \end{cases}$

One has $u_\delta = u_0 - \int_D g h dy + \int_D g(x, y)(h - h_\delta) dy$,

$$|u_\delta(x)| \geq |u(x)| - c \int_{N_\delta} \frac{dy}{4\pi|x - y|} \geq \delta - I(\delta), \quad x \in D_\delta, \quad c = \max_{x \in N_\delta} |h(x)|.$$

If one proves, that $I(\delta) = o(\delta)$, $\delta \to 0$, $\forall x \in D_\delta$ then $q_\delta \in L^\infty(D)$, and Claim 1 is proved. □

Claim 2.

$$I(\delta) = \mathcal{O}(\delta^2 |\ln(\delta)|), \quad \delta \to 0.$$

Proof of Claim 2.

$$\int_{N_\delta} \frac{dy}{|x - y|} \leq \int_{N_\delta} \frac{dy}{|y|} = c_1 \int_0^{c_2\delta} \rho \int_0^1 \frac{ds_3}{\sqrt{\rho^2 + s_3^2}} d\rho$$

$$= c_1 \int_0^{c_2\delta} d\rho \rho \ln(s_3 + \sqrt{\rho^2 + s_3^2})|_0^1 \leq c_3 \int_0^{c_2\delta} \rho \ln\left(\frac{1}{\rho}\right) d\rho$$

$$\leq \mathcal{O}(\delta^2 |\ln(\delta)|).$$

The condition $|\nabla u_j|_l \geq c > 0, j = 1, 2$, implies that a tubular neighborhood of the line l, $N_\delta = \{x : \sqrt{|u_1|^2 + |u_2|^2} \leq \delta\}$, is included in a region $\{x : |x| \leq c_2\delta\}$ and includes a region $\{x : |x| \leq c_2'\delta\}$. This follows from the estimates

$$c_2'\rho \leq |u(x)| = |\nabla u(\xi) \cdot (x - \xi)| \leq c_2\rho.$$

Here $\xi \in l$, x is a point on a plane passing through ξ and orthogonal to l, $\rho = |x - \xi|$, and $\delta > 0$ is sufficiently small, so that the terms of order ρ^2 are negligible, $c_2 = \max_{\xi \in l} |\nabla u(\xi)|$, $c_2' = \min_{\xi \in l} |\nabla u(\xi)|$.

Claim 2, and, therefore, Lemma 4 are proved.   □

Therefore, Theorem 2 is proved.   □

Let us describe a numerical method for calculation of h given $f(\beta)$ and $\epsilon > 0$. Let $\{\phi_j\}$ be a basis in $L^2(D)$, $h_n = \sum_{j=1}^{n} c_j^{(n)} \phi_j$, $\psi_j(\beta) := -\frac{1}{4\pi} \int_D e^{-ik\beta \cdot x} \phi_j(x) dx$. Consider the problem:

$$\left\| f(\beta) - \sum_{j=1}^{n} c_j^{(n)} \psi_j(\beta) \right\| = \min. \tag{24}$$

A necessary condition for (24) is a linear algebraic system for $c_j^{(n)}$.

References

[1]   H. C. Hulst van de, *Light scattering by small particles*, Dover, New York, 1961.

[2]   L. Landau, L. Lifschitz, *Electrodynamics of continuous media*, Pergamon Press, Oxford, 1984.

[3]   P. Martin, *Multiple scattering*, Cambridge Univ. Press, Cambridge, 2006.

[4]   A. G. Ramm, *Scattering by obstacles*, D. Reidel, Dordrecht, 1986.

[5]   A. G. Ramm, *Wave scattering by small bodies of arbitrary shapes*, World Sci. Publishers, Singapore, 2005.

[6]   A. G. Ramm, Scattering by many small bodies and applications to condensed matter physics, *Europ. Phys. Lett.*, **80**, 44001, (2007).

[7]   A. G. Ramm, Many-body wave scattering by small bodies and applications, *J. Math. Phys.*, **48**, 103511, (2007).

[8]   A. G. Ramm, Wave scattering by small particles in a medium, *Phys. Lett. A*, **367**, 156–161, (2007).

[9]   A. G. Ramm, Wave scattering by small impedance particles in a medium, *Phys. Lett. A*, **368**, 164–172, (2007).

[10] A. G. Ramm, Distribution of particles which produces a desired radiation pattern, *Communic. in Nonlinear Sci. and Numer. Simulation*, **12**, N7, (2007), 1115–1119.

[11] A. G. Ramm, Distribution of particles which produces a "smart" material, *Jour. Stat. Phys.*, **127**, 915–934, (2007).

[12]   A. G. Ramm, Distribution of particles which produces a desired radiation pattern, *Physica B*, **394**, 253–255, (2007).

[13]   A. G. Ramm, Creating wave-focusing materials, *LAJSS (Latin-American Journ. of Solids and Structures)*, **5**, 119–127, (2008).

[14]   A. G. Ramm, Electromagnetic wave scattering by small bodies, *Phys. Lett. A*, **372**, 4298–4306, (2008).

[15]   A. G. Ramm, Wave scattering by many small particles embedded in a medium, *Phys. Lett. A*, **372**, 3064–3070, (2008).

[16]   A. G. Ramm, Preparing materials with a desired refraction coefficient and applications, *In the book "Topics in Chaotic Systems: Selected Papers from Chaos 2008 International Conference"*, Editors C. Skiadas, I. Dimotikalis, Char. Skiadas, World Sci. Publishing, pp. 265–273, (2009),

[17]   A. G. Ramm, Preparing materials with a desired refraction coefficient,*Nonlinear Analysis: Theory, Methods and Appl.*, **70**, e186-e190, (2009).

[18]   A. G. Ramm, Creating desired potentials by embedding small inhomogeneities, *J. Math. Phys.*, **50**, 123525, (2009).

[19]   A. G. Ramm, A method for creating materials with a desired refraction coefficient, *Internat. Journ. Mod. Phys B*, **24**, 5261–5268, (2010).

[20]   A. G. Ramm, Materials with a desired refraction coefficient can be created by embedding small particles into the given material, *International Journal of Structural Changes in Solids (IJSCS)*, **2**, 17–23, (2010).

[21]   A. G. Ramm, Wave scattering by many small bodies and creating materials with a desired refraction coefficient, *Afrika Matematika*, **22**, 33–55, (2011).

[22]   A. G. Ramm, Scattering by many small inhomogeneities and applications, *In the book "Topics in Chaotic Systems: Selected Papers from Chaos 2010 International Conference"*, Editors C. Skiadas, I. Dimotikalis, Char. Skiadas, World Sci.Publishing, pp. 41–52, (2011),

[23]   A. G. Ramm, Collocation method for solving some integral equations of estimation theory, *Internat. Journ. of Pure and Appl. Math.*, **62**, 57–65, (2010).

[24]   A. G. Ramm, A method for creating materials with a desired refraction coefficient, *Internat. Journ. Mod. Phys B*, **24**, 27, (2010), 5261–5268.

[25]   A. G. Ramm, Electromagnetic wave scattering by a small impedance particle of arbitrary shape, *Optics Communications*, **284**, (2011), 3872–3877.

[26] A. G. Ramm, Scattering of scalar waves by many small particles, *AIP Advances*, **1**, 022135, (2011).

[27] A. G. Ramm, Scattering of electromagnetic waves by many thin cylinders, *Results in Physics*, **1**, N1, (2011), 13–16.

[28] A. G. Ramm, Electromagnetic wave scattering by many small perfectly conducting particles of an arbitrary shape, *Optics Communications*, **285**, N18, (2012), 3679–3683.

[29] A. G. Ramm, Electromagnetic wave scattering by small impedance particles of an arbitrary shape, *J. Appl. Math and Comput., (JAMC)*, **43**, N1, (2013), 427–444.

[30] A. G. Ramm, Many-body wave scattering problems in the case of small scatterers, *J. of Appl. Math and Comput., (JAMC)*, **41**, N1, (2013), 473–500.

[31] A. G. Ramm, Scattering of Acoustic and Electromagnetic *Waves by Small Bodies of Arbitrary Shapes. Applications to Creating New Engineered Materials*, Momentum Press, New York, 2013.

[32] A. G. Ramm, *Creating materials with a desired refraction coefficient*, IOP Publishers, Bristol, UK, 2020 (Second edition).

[33] A. G. Ramm, How can one create a material with a prescribed refraction coefficient? *Sun Text Review of Material Science*, **1:1**, (2020), 102.

[34] A. G. Ramm, *Scattering by obstacles and potentials*, World Sci. Publ., Singapore, 2017.

[35] A. G. Ramm, N. Tran, A fast algorithm for solving scalar wave scattering problem by billions of particles, *Journal of Algorithms and Optimization*, **3**, N1, (2015), 1–13. Open access: http://www.academ icpub.org/jao/Issue.aspx?Abstr=false.

[36] J. Rayleigh, *Scientific papers*, Cambridge Univ. Press, Cambridge, 1992.

Appendix D: Is Creating Materials with a Desired Refraction Coefficient Practically Possible?*

Alexander G. Ramm

Department of Mathematics
Kansas State University, Manhattan, USA
ramm@ksu.edu

Abstract

A theory of many-body wave scattering is developed under the assumption $a \ll d \ll \lambda$, where a is the characteristic size of the small body, d is the distance between neighboring bodies and λ is the wavelength in the medium in which the bodies are embedded. The multiple scattering is essential under these assumptions. The author's theory is used for c reating materials with a desired refraction coefficient. This theory can be used in practice. A recipe for creating materials with a desired refraction coefficient is formulated. Materials with a desired radiation pattern, for example, wave-focusing materials, can be created.

Keywords: wave scattering by many small bodies; smart materials.

---

*This article is published in "Characterization and application of nanomaterials". *Bioinorganic Chemistry and Applications*, **6**(1), 1–5 (2023).

1   Introduction

The aim of this paper is to give an affirmative answer to the question in the title of this paper. This brings potentially many possibilities for progress in technology.

There is a large literature on wave scattering by small bodies, starting from Rayleigh's work (1871), [42], [3], [4]. If the scatterer is small then the scattered field can be calculated analytically for bodies of arbitrary shapes, see [7].

The many-body wave scattering problem was discussed in the literature mostly numerically, if the number of scatterers was small, or under the assumption that the influence of the waves, scattered by other particles on a particular particle is negligible, [5]. This corresponds to the case when *the distance d between neighboring particles is much larger than the wavelength λ, and the characteristic size a of a small body (particle) is much smaller than λ.* Theoretically and practically the assumptions

$$a \ll \lambda, \quad d \gg \lambda, \tag{1}$$

are the simplest ones which allow one *to neglect multiple scattering.* By $k = \frac{2\pi}{\lambda}$ the wave number is denoted.

In the author's theory the basic assumptions are

$$a \ll d \ll \lambda, \tag{2}$$

and *the multiple scattering is of basic importance under these assumptions,* [6]–[37]. It is clear that assumption (2) can be practically realized. Its importance comes from the fact that the author gave a rigorous asymptotically exact solution of the many-body scattering problem under assumption (2) when $a \to 0$. This solution can be well approximated numerically by the particles of the size $a > 30nm$. Practically the size of a can be found by comparison of the solution for some a and for $\frac{a}{2}$. If these solutions are practically close, then one considers this a as suitable. The aim of this paper is to show that our theory can be used practically.

In reference [38] for the first time the author's theory was used for solving the scattering problem for 10 billions small particles. This problem was solved numerically and numerical results were presented.

Let us formulate the wave scattering problems we deal with. Let D be a bounded domain in $\mathbb{R}^3$ with a sufficiently smooth boundary. The scattering problem consists of finding the solution to the problem:

$$(\nabla^2 + k^2)u = 0 \text{ in } G' := \mathbb{R}^3 \setminus G, \quad G := \cup_{m=1}^M D_m, \quad k = \text{const} > 0, \tag{3}$$

where $D_m = B(x_m, a)$ is an impedance ball, centered at x_m and of small radius a,

$$u = u_0 + v, \quad u_0 = e^{ik\alpha \cdot x}, \quad \alpha \in S^2, \tag{4}$$

S^2 is the unit sphere in $\mathbb{R}^3$, u_0 is the incident field, v is the scattered field satisfying the radiation condition

$$v_r - ikv = o\left(\frac{1}{r}\right), \quad r := |x| \to \infty, \ v_r := \frac{\partial v}{\partial r}, \tag{5}$$

and u satisfies the impedance boundary condition (bc) on the boundary of G:

$$u_N - \zeta_m u = 0, \quad \text{on } S_m, \quad \text{Im}\zeta_m \le 0, \tag{6}$$

where ζ_m is a constant, N is the unit normal to $S := \cup_{m=1}^M S_m$, pointing out of $G := \cup_{m=1}^M D_m$, and S_m is the surface of $D_m = B(x_m, a)$.

By refraction coefficient $n(x)$ the coefficient in the equation

$$(\nabla^2 + k^2 n^2(x))u = (\nabla^2 + k^2 - q(x))u = 0 \tag{7}$$

is understood, where $q(x) := k^2(n^2(x) - 1)$.

Let $g(x, y) = \frac{e^{ik|x-y|}}{4\pi|x-y|}$. Then $(\nabla^2 + k^2)g(x, y) = -\delta(x - y)$, where $\delta(x)$ is the delta function.

Let us distribute small impedance particles $D_m = B(x_m, a)$ in D so that

$$\mathbb{N}(\Delta) = a^{\kappa-2}|\Delta|[1 + o(1)], \quad a \to 0, \tag{8}$$

where $\Delta \subset D$ is an arbitrary connected open subset of D, $|\Delta|$ is its volume, $\kappa \in (0, 1)$ is a number the experimenter may choose arbitrarily and $\mathbb{N}(\Delta)$ is the number of particles in Δ. Throughout this paper the important assumptions $a \ll d \ll \lambda$ and (8) are satisfied.

As $a \to 0$ the number of small particles $\mathbb{N}(\Delta)$ in (8) tends to infinity since $\kappa - 2 < 0$.

We assume in this paper (for simplicity only) that the small particles are distributed in the domain D and the refraction coefficient in D equals to 1. In the monograph [33] it is assumed that D is filled with the material whose refraction coefficient $n_0(x)$ is known and we wanted to create in D the material with the desired refraction coefficient $n(x)$.

The boundary impedances ζ_m are chosen by the formula

$$\zeta_m = a^{-\kappa} h(x_m), \tag{9}$$

where $h(x)$ is a continuous function in D, $\mathrm{Im}\, h \leq 0$.

It will be clear from Section 3 that the function $h(x)$ can be determined by choosing a suitable boundary impedance $\zeta(x)$. When $a \to 0$, the ζ_m and $h(x_m)$ can be considered as continuous functions $\zeta(x)$ and $h(x)$.

2   Solution of many-body scattering problem

We look for the solution of the form

$$u = u_0 + \sum_{m=1}^{M} \int_{S_m} g(x,s)\sigma_m(s)ds = \sum_{m=1}^{M} g(x,x_m)Q_m + J, \tag{10}$$

where $\sigma_m(s)$ are unknown, $Q_m := \int_{S_m} \sigma_m(s)ds$. One may think about σ_m as of charge densities on S_m and of Q_m as of total charge on the surface S_m. We prove that

$$J := \sum_{m=1}^{M} \int_{S_m} [g(x,s) - g(x,x_m)]\sigma_m(s)ds \tag{11}$$

is negligible compared to

$$I := \sum_{m=1}^{M} g(x,x_m)Q_m, \tag{12}$$

so

$$J \ll I \quad \text{as } a \to 0. \tag{13}$$

We prove that the field u satisfies the following integral equation as $a \to 0$:

$$u(x) = u_0(x) - 4\pi \int_D g(x,y)h(y)u(y)dy, \tag{14}$$

where $h(x_m) = \frac{\zeta_m}{a^\kappa}$, and, since there are sufficiently many points $x_m \in D$, the function $h(x)$ is uniquely determined in D if the boundary impedances are known.

Apply the operator $\nabla^2 + k^2$ to both sides of equation (14) and get

$$(\nabla^2 + k^2 - 4\pi h(x))u(x) := (\nabla^2 + k^2 n^2(x))u(x) = 0. \tag{15}$$

Therefore,

$$n^2(x) = 1 - 4\pi k^{-2}h(x). \tag{16}$$

We omit details since they can be found in the author's publications listed in the References, in particular, in monograph [33].

If originally in D were material with the known refraction coefficient $n_0(x)$, then formula (16) were $n^2(x) = n_0^2(x) - 4\pi h(x)N(x)k^{-2}$, where $N(x)$ is the distribution density for the small particles, see [33]. In this paper we assume (for simplicity only) that $N(x) = 1$, see formula (8).

3 Recipe for creating materials with a desired refraction coefficient

Let us formulate a recipe for creating materials with a desired refraction coefficient. Formula (16) shows that if $h(x)$ is chosen properly, then any $n(x)$ can be obtained in D.

Recipe for creating materials with a desired refraction coefficient:

(a) *Calculate by formula (16) the function $h(x)$;*
(b) *Distribute small impedance balls in the domain D by the distribution law (8). The boundary impedances of these balls are defined by the function $h(x)$.*

Theorem 1. *The refraction coefficient of the resulting medium tends to the desired coefficient $n(x)$ as $a \to 0$.*

Let us show that practically negative refraction coefficient $n(x)$ can be obtained by the above recipe. Denote $b := 4\pi k^{-2} > 0$ and write equation (16) as

$$n(x) = (1 - bh(x))^{1/2} = |1 - bh(x)|^{1/2} e^{\phi/2}, \qquad (17)$$

where ϕ is the argument of $1 - bh(x)$. Since the operator in (14) is of Fredholm type, it remains Fredholm type under small perturbations. Therefore one can take $h - i\epsilon$, where $\epsilon > 0$ is sufficiently small, and equation (14) will still have a unique solution.

By choosing h so that $\text{Re}(1 - bh) > 0$ and $\text{Im}(1 - bh) < 0$ and small, one gets the argument $\phi = 2\pi - \delta$, where $\delta > 0$ is arbitrarily small if ϵ is sufficiently small. Then $n(x)$ will be nearly negative: its argument will be $\pi - \delta/2$.

4   Creating materials with a desired radiation pattern

Let us define what we mean by the radiation pattern. Consider the scattering problem for the equation:

$$\nabla^2 u + k^2 u - q(x)u = 0, \quad u = e^{ik\alpha \cdot x} + v, \qquad (18)$$

where v satisfies the radiation condition. Assume that $k > 0$ and $\alpha \in S^2$ are fixed. Then the scattering amplitude $A(\beta, \alpha, k) = A(\beta)$, where the dependence on k, α is dropped since k and α are fixed. The formula for the scattering amplitude is known, see, e.g., [37]:

$$A(\beta) := A_q(\beta) = -\frac{1}{4\pi} \int e^{-ik\beta \cdot y} q(y) u(y) dy. \qquad (19)$$

We call $A(\beta)$ the radiation pattern.

Consider an inverse problem (IP):

Given an arbitrary $f(\beta) \in L^2(S^2)$ and an arbitrary small $\epsilon > 0$, can one find a $q \in L^2(D)$ such that

$$\|f(\beta) - A_q(\beta)\|_{L^2(S^2)} < \epsilon. \qquad (20)$$

This inverse problem was not formulated and was not studied in the works of other authors, to our knowledge.

Our result is stated in Theorem 2.

Theorem 2. *For any $f(\beta) \in L^2(S^2)$ and an arbitrary small $\epsilon > 0$ there is a $q \in L^2(D)$ such that (20) holds.*

Since small perturbations of q result in small perturbations of $A(\beta)$, there are infinitely many potentials q for which inequality (20) holds.

The conclusion of Theorem 2 follows from Lemmas 3 and 4.

Lemma 3. *The set $\{\int_D e^{-ik\beta \cdot x} h(x) dx\}_{\forall h \in L^2(D)}$ is dense in $L^2(S^2)$.*

Corollary 1. *Given $f \in L^2(S^2)$ and $\epsilon > 0$, one can find $h \in L^2(D)$ such that*

$$\|f(\beta) + \frac{1}{4\pi} \int_D e^{-ik\beta \cdot x} h(x) dx\| < \epsilon.$$

Lemma 4. *The set $\{q(x)u(x, \alpha)\}_{\forall q \in L^2(D)}$ is dense in $L^2(D)$.*

Corollary 2. *Given $h \in L^2(D)$ and $\epsilon > 0$, one can find $q \in L^2(D)$ such that*

$$\|h(x) - q(x)u(x, \alpha)\|_{L^2(D)} < \epsilon.$$

Since the scattering amplitude

$$A(\beta) = -\frac{1}{4\pi} \int_D e^{-ik\beta \cdot x} h(x) dx$$

depends continuously on h, the inverse problem **IP** *is solved by Lemmas 3 and 4.*

Proofs are omitted. They can be found in [33].

5   Discussion

How is the theory, outlined in the previous sections, can be used practically?

To create a material with a desired refraction coefficient, or a material with a refraction coefficient close to the desired, is practically very important. To my knowledge, there were no general methods for creating material with a desired refraction coefficient.

To use the theory, outlined in this paper and in the monographs [33], [34], [35], one has to solve a technological problem: how to prepare a small particle, say, a ball of radius a, with the prescribed boundary impedance ζ. This problem should be solvable, see [35] for arguments supporting this conclusions. If this technological problem is solved, then the recipe outlined in this paper (and in the author's monographs [33], [34], [35] can be immediately used in practice.

The problem of creating materials with a desired radiation pattern, the wave focusing materials, for example, was not investigated earlier. This problem is of great practical interest. The usual bodies scatter waves mostly backwards, somewhat sidewise and a little forwards. If one creates a body which scatters waves, for example, in a given solid angle, this would be of great practical interest. Such a body can be created as follows from the theory outlined in the previous section.

The author wrote this paper in an attempt to draw attention of the specialists in material sciences to the theory he has developed for creating materials with the desired refraction coefficient.

The author is not aware of the experimental results based on his theory. Such results are very desirable. There are numerical results, based on his theory, see [1], [2].

References

[1] Andriychuk, M., Ramm, A. G., Numerical solution of many-body wave scattering problem for small particles and creating materials with desired refraction coefficient, chapter in the book: "Numerical Simulations of Physical and Engineering Processes", InTech., Vienna, 2011, pp. 1–28. (edited by Jan Awrejcewicz) ISBN 978-953-307-620-1

[2] Andriychuk, M., Ramm, A. G., Scattering of electromagnetic waves by many thin cylinders: theory and computational modeling, Optics Communications, 285, N20, (2012), 4019–4026.

[3] H. C. Hulst van de, *Light scattering by small particles*, Dover, New York, 1961.

[4] L. Landau, L. Lifschitz, *Electrodynamics of continuous media*, Pergamon Press, Oxford, 1984.

[5] P. Martin, *Multiple scattering*, Cambridge Univ. Press, Cambridge, 2006.

[6] A. G. Ramm, *Scattering by obstacles*, D.Reidel, Dordrecht, 1986.

[7] A. G. Ramm, *Wave scattering by small bodies of arbitrary shapes*, World Sci. Publishers, Singapore, 2005.

[8] A. G. Ramm, Scattering by many small bodies and applications to condensed matter physics, *Europ. Phys. Lett.*, **80**, 44001, (2007).

[9] A. G. Ramm, Many-body wave scattering by small bodies and applications, *J. Math. Phys.*, **48**, 103511, (2007).

[10] A. G. Ramm, Wave scattering by small particles in a medium, *Phys. Lett. A*, **367**, 156–161, (2007).

[11] A. G. Ramm, Wave scattering by small impedance particles in a medium, *Phys. Lett. A*, **368**, 164–172, (2007).

[12] A. G. Ramm, Distribution of particles which produces a desired radiation pattern, *Communic. in Nonlinear Sci. and Numer. Simulation*, **12**, N7, (2007), 1115–1119.

[13] A. G. Ramm, Distribution of particles which produces a "smart" material, *Jour. Stat. Phys.*, **127**, 915–934, (2007).

[14] A. G. Ramm, Distribution of particles which produces a desired radiation pattern, *Physica B*, **394**, 253–255, (2007).

[15] A. G. Ramm, Creating wave-focusing materials, *LAJSS (Latin-American Journ. of Solids and Structures)*, **5**, 119–127, (2008).

[16] A. G. Ramm, Electromagnetic wave scattering by small bodies, *Phys. Lett. A*, **372**, 4298–4306, (2008).

[17] A. G. Ramm, Wave scattering by many small particles embedded in a medium, *Phys. Lett. A*, **372**, 3064–3070, (2008).

[18] A. G. Ramm, Preparing materials with a desired refraction coefficient and applications, *In the book "Topics in Chaotic Systems: Selected Papers from Chaos 2008 International Conference"*, Editors C. Skiadas, I. Dimotikalis, Char. Skiadas, World Sci. Publishing, pp. 265–273, (2009).

[19] A. G. Ramm, Preparing materials with a desired refraction coefficient, *Nonlinear Analysis: Theory, Methods and Appl.*, **70**, e186-e190, (2009).

[20] A. G. Ramm, Creating desired potentials by embedding small inhomogeneities, *J. Math. Phys.*, **50**, 123525, (2009).

[21] A. G. Ramm, A method for creating materials with a desired refraction coefficient, *Internat. Journ. Mod. Phys B*, **24**, 5261–5268, (2010).

[22] A. G. Ramm, Materials with a desired refraction coefficient can be created by embedding small particles into the given material, *International Journal of Structural Changes in Solids (IJSCS)*, **2**, 17–23, (2010).

[23] A. G. Ramm, Wave scattering by many small bodies and creating materials with a desired refraction coefficient, *Afrika Matematika*, **22**, 33–55, (2011).

[24] A. G. Ramm, Scattering by many small inhomogeneities and applications, *In the book "Topics in Chaotic Systems: Selected Papers from Chaos 2010 International Conference"*, Editors C. Skiadas, I. Dimotikalis, Char. Skiadas, World Sci. Publishing, pp. 41–52, (2011).

[25] A. G. Ramm, Collocation method for solving some integral equations of estimation theory, *Internat. Journ. of Pure and Appl. Math.*, **62**, 57–65, (2010).

[26] A. G. Ramm, A method for creating materials with a desired refraction coefficient, *Internat. Journ. Mod. Phys B*, **24**, 27, (2010), 5261–5268.

[27] A. G. Ramm, Electromagnetic wave scattering by a small impedance particle of arbitrary shape, *Optics Communications*, **284**, (2011), 3872–3877.

[28] A. G. Ramm, Scattering of scalar waves by many small particles, *AIP Advances*, **1**, 022135, (2011).

[29] A. G. Ramm, Scattering of electromagnetic waves by many thin cylinders, *Results in Physics*, **1**, N1, (2011), 13–16.

[30] A. G. Ramm, Electromagnetic wave scattering by many small perfectly conducting particles of an arbitrary shape, *Optics Communications*, **285**, N18, (2012), 3679–3683.

[31] A. G. Ramm, Electromagnetic wave scattering by small impedance particles of an arbitrary shape, *J. Appl. Math and Comput., (JAMC)*, **43**, N1, (2013), 427–444.

[32] A. G. Ramm, Many-body wave scattering problems in the case of small scatterers, *J. of Appl. Math and Comput., (JAMC)*, **41**, N1, (2013), 473–500.

[33] A. G. Ramm, Scattering of Acoustic and Electromagnetic *Waves by Small Bodies of Arbitrary Shapes. Applications to Creating New Engineered Materials*, Momentum Press, New York, 2013.

[34] A. G. Ramm, *Creating materials with a desired refraction coefficient*, IOP Concise Physics, Mprgan and Claypool Publishers, San Rafael, California, 2017.

[35] A. G. Ramm, *Creating materials with a desired refraction coefficient*, IOP Publishers, Bristol, UK, 2020 (Second edition).

[36] A. G. Ramm, How can one create a material with a prescribed refraction coefficient? *Sun Text Review of Material Science*, **1:1**, (2020), 102.

[37] A. G. Ramm, *Scattering by obstacles and potentials*, World Sci. Publ., Singapore, 2017.

[38]   A. G. Ramm, N. Tran, A fast algorithm for solving scalar wave scattering problem by billions of particles, *Jour. of Algorithms and Optimization*, **3**, N1, (2015), 1–13. Open access: http://www.academic pub.org/jao/Issue.aspx?Abstr=false.

[39]   A. G. Ramm, Cong Tuan Son Van, Creating materials in which heat propagates along a line: theory and numerical results, Pure and Applied Functional Analysis, (PAFA), 2, N4, (2017), 639–648. Open access Journal.

[40]   A. G. Ramm, Finding a method for producing small impedance particles with prescribed boundary impedance is important, J. Phys. Res. Appl., 1:1, (2017), 1–3. Open access Journal.

[41]   A. G. Ramm, Wave scattering by many small impedance particles and applications, Reports on Mathem. Phys. (ROMP), 90, N2, (2022), 193–202.

[42]   J. Rayleigh, *Scientific papers*, Cambridge Univ. Press, Cambridge, 1992.

Bibliographical Notes

Wave scattering by small bodies was discussed by Rayleigh in 1871 and by many authors later. Rayleigh understood that the main term in the scattered field is the dipole radiation, and this field is $O(a^3)$, where a is the characteristic size of the small body. The scattered field is proportional to the induced dipole moment P, and this moment is proportional to the volume V of the small scatterer, $V = O(a^3)$, to the value of the incident field E at the point where the small body is, and to the polarization tensor α_{jm}. Thus, $P_j = \alpha_{jm}\epsilon_0 V E_m$, where the summation is understood over the repeated indices.

Rayleigh did not give formulas for calculating the tensor α_{jm} for bodies of an arbitrary shape. This was done about 100 years later in Ramm (1969a, 1969b, 1970, 1971b, 1971c, 1974a, 1974b, 1980a, 1980b, 1982, 1986, 2005a, 2005b).

It turns out that the basic integral equations of the static problem are on the smallest characteristic value of the integral operators in these equations, that this value is a simple pole of the resolvent of the operator of these integral equations, and the free term in these equations satisfies the orthogonality conditions necessary and sufficient for the solvability of these equations. These equations arise in the problems of electrostatics, magnetostatics, elasticity theory, etc.

The author constructed convergent iterative methods for solving basic equations for the aforementioned static fields and studied stability of the iterative process and of the solutions to the integral equations under the above conditions, see Ramm (1980b, 1982, 2005b).

In Chapters 1–4 of this book, the author develops wave scattering theory in the case of one and many small impedance particles of an arbitrary shape. Analytical formulas are derived for the field, scattered by one small impedance particle of an arbitrary shape. Many-body wave scattering problem is solved analytically and numerically in the case of small impedance particles of an arbitrary shape. The central idea is to reduce this problem to finding some numbers rather than unknown boundary functions.

The basic assumptions are: smallness of the particles: $ka \ll 1$, the distance d between closest neighboring particles is much larger than $a, d \gg a$. However, $d \ll \lambda$ is allowed, so the multiple scattering effects are crucial. Equation for the effective field is derived in the limiting medium when the size a of the particles tends to zero and their number tends to infinity at the rate $O(\frac{1}{a^{2-\kappa}}), 0 \le \kappa < 1$, where κ is a given parameter that can be chosen by the experimenter.

This theory allows the author to formulate a recipe for creating materials with a desired refraction coefficient.

This coefficient can be created negative, which is of interest in the theory of metamaterials.

This coefficient can be created so that the resulting material has a desired radiation pattern.

These results are taken from the papers Ramm (2007f, 2007g, 2007i, 2008a, 2008c, 2009a, 2009d, 2009e, 2010b, 2011d, 2013c, 2014a).

The field, scattered by one small impedance particle with boundary impedance $\zeta = \frac{h}{a^{\kappa}}$, is of the order $O(a^{2-\kappa})$, which is much larger than $O(a^3)$. It would be interesting to use this practically.

Besides the impedance boundary condition other boundary conditions are investigated in Chapters 1–3. These include the Dirichlet boundary condition, the Neumann boundary condition, and the transmission boundary conditions. In the case of the Neumann and transmission boundary conditions, the equation for the effective field cannot be reduced to a local differential equation.

Numerical implementation of our theory is given in Andriychuk and Ramm (2010), Andriychuk and Ramm (2011) and in Indratno and Ramm (2010).

In Chapters 5–7, the theory developed in Chapters 1–4 is generalized to the case of electromagnetic (EM) wave scattering by one and many small impedance particles of an arbitrary shape. An analytical formula is obtained for the EM field scattered by one small impedance particle of an arbitrary shape. This field is of the order $O(a^{2-\kappa})$, which is much larger than $O(a^3)$ when $a \to 0$. Analytical formulas are obtained and numerical methods are developed for solving many-body EM wave scattering problems in the case of small impedance particles of an arbitrary shape. The physical assumptions are the same as in the case of scalar wave scattering. The theory, developed in these chapters, is applied to the problem of creating materials with a desired refraction coefficient and a desired magnetic permeability.

The results of these chapters are partly new and partly published in Ramm (2013a).

In Chapter 8, the problem of EM wave scattering by many thin impedance cylinders (nanowires) is studied. The problem is solved analytically and an equation for the limiting effective field is derived as $a \to 0$ and the number of thin cylinders tends to infinity.

It is shown that the refraction coefficient in the limiting medium differs from the original one. The new refraction coefficient is calculated analytically. It is shown how to choose boundary impedances of the cylinders in order to create a desired refraction coefficient of the limiting medium. The results of this chapter are based on Ramm (2013d).

Chapter 9 deals with the heat transfer in the medium where many small particles are embedded. The results of this chapter are taken from Ramm (2013b).

Chapter 10 deals with wave scattering by many potentials with a small support. Again, an analytical formula is derived for the effective field in the limiting medium. The result is taken from Ramm (2010c).

Chapter 11 contains some known results of potential theory in Section 11.1. The derivation of these results is self-contained and allows the reader to use this book without using other books. Section 11.2 contains new material: it is shown in this section when potentials of the single layer can be expressed as potentials of double layer and vice versa. The ideas are similar to the ideas from Ramm (1986), pp. 71–83, but the presentation is different. Section 11.3 deals with the asymptotic behavior of the solution to the Helmholtz equation when the boundary impedance tends to infinity. The result is similar to the one in Alber and Ramm (2009), but the presentation is different. Section 11.4 presents some properties of electrical capacitance. The result is taken from Ramm (2012, 2013e).

Chapter 12 presents a version of the collocation method used in Chapters 1–7. Convergence of this method for solving integral equation for the effective field is proved and it is shown how this result can be used for a justification of a version of theory. There are many books on the theory, see Bensoussan *et al.* (2011), Marčenko and Khruslov (2006), Zhikov *et al.* (1994), to mention a few. Our version of the theory differs in several respects: no periodicity assumptions are made, no assumption about discreteness of the spectrum of the operator in a periodic cell is made, our operators are non-self-adjoint and have continuous spectrum. The results are based on Ramm (2009a, 2011d).

In Chapter 13, some inverse problems are briefly discussed. These problems are based on the theory presented in Chapters 1–7. The first problem, considered in Section 13.1, is the following: there is a large body D_1 (say, a ship) wave scattering on this body is known; there is a small body D_2 (say, a mine) in the far zone of D_1. One observes the field scattered by both bodies $D_1 \cup D_2$ and wants to calculate the position of D_2 and estimate its size. A similar problem is treated in Olshansky and Ramm (2009). In Section 13.2, the problem of finding the positions and number of small subsurface inhomogeneities from the scattering data measured on the surface for various positions of the source of the waves is solved. This problem can be considered as a model for ultrasound mammography. The result is taken from Ramm (2000). Numerical results, based on the developed method, are presented in Gutman and Ramm (2000). In Section 13.3, an inverse radio measurements problem is solved. The problem consists in an accurate measurement of a complicated distribution of electromagnetic field in an aperture of a mirror antenna. Assume that the wavelength is $\lambda = 3$ cm, that is, radio wave diapason. There is a small probe $a \ll \lambda$, say $a = 0.1$ cm, which is moved along the aperture by a mechanical device transparent for the waves at $\lambda = 3$ cm. The EM field E, H scattered by the small probe is measured. From these measurements, one wants to calculate the value E_0, H_0 of the field at the point where the probe is, which existed before the probe was placed at this point. The shape and electromagnetic parameters of the probe are assumed known.

This problem has been solved in Ramm (2005b), Section 7.6, and we borrow the method of the solution from there.

In Appendix A, some definitions and results from functional analysis are briefly stated for convenience of the reader. These results are known. The presentation in Section A.3 follows Ramm (2001, 2003).

Appendices B,C,D are the author's papers published after the publication of the monograph Ramm (2013g).

The monographs Ramm (2020) and Ramm (2017) treat concisely the many-body wave scattering problem and creating materials with a desired refraction coefficient. In Ramm (2017), the many-body scattering problem is solved under the assumption that the small scatterers are located inside a bounded, closed, perfectly conducting surface.

Bibliography

Alber, H.-D. and A. G. Ramm (2009). Asymptotics of the solution to Robin problem. *Journal of Mathematical Analysis and Applications 349*(1), 156–164.

Andriychuk, M. and A. G. Ramm (2010). Scattering by many small particles and creating materials with a desired refraction coefficient. *International Journal of Computing Science and Mathematics 3*(1), 102–121.

Andriychuk, M. and A. G. Ramm (2011). Numerical solution of many-body wave scattering problem for small particles and creating materials with desired refraction coefficient. In J. Awrejcewicz (Ed.), *Numerical Simulations of Physical and Engineering Processes*, pp. 1–28. InTech., Vienna.

Andriychuk, M., S. Indratno, and A. G. Ramm (2012). Electromagnetic wave scattering by a small impedance particle: Theory and modeling. *Optics Communications 285*(7), 1684–1691.

Bensoussan, A., J.-L. Lions, and G. Papanicolaou (2011). *Asymptotic Analysis for Periodic Structures*, Volume 374. American Mathematical Society, Providence, RI.

Gilbarg, D. and N. S. Trudinger (1983). *Elliptic Partial Differential Equations of Second Order*, Volume 224. Springer Verlag, Berlin.

Günter, N. (1967). *Potential Theory, and Its Applications to Basic Problems of Mathematical Physics*. Ungar, New York.

Gutman, S. and A. G. Ramm (2000). Application of the hybrid stochastic-deterministic minimization method to a surface data inverse scattering problem. *Operator Theory and Its Applications 25*, 293.

Hörmander, L. (1983–1985). *The Analysis of Linear Partial Differential Operators*, Volumes I–IV. Springer-Verlag, Berlin.

Indratno, S. W. and A. G. Ramm (2010). Creating materials with a desired refraction coefficient: Numerical experiments. *International Journal of Computing Science and Mathematics* 3(1), 76–101.

Kantorovich, L. V. and G. P. Akilov (1982). *Functional Analysis.* Pergamon Press, Oxford.

Kato, T. (1984). *Perturbation Theory for Linear Operators.* Springer Verlag, Berlin.

Landau, L. D. and E. M. Lifshitz (1984). *Electrodynamics of Continuous Medium.* Pergamon Press, Oxford.

Lebedev, N. N. (1972). *Special Functions and Their Applications.* Dover, New York.

Marčenko, V. A. and E. Y. Khruslov (2006). *Homogenization of Partial Differential Equations.* Birkhäser, Boston.

Mie, G. (1908). Beiträge zur Optik trüber Medien, speziell kolloidaler Metallösungen. *Annalen der Physik 25*, 377–445.

Mikhlin, S. G. and S. Prössdorf (1986). *Singular Integral Operators.* Springer-Verlag, Berlin.

Olshansky, Y. and A. G. Ramm (2009). Finding the position of a small body in the presence of a large body from the scattering data. *Inverse Problems in Science and Engineering 17*(5), 699–712.

Ramm, A. G. (1969a). Calculation of the scattering amplitude for the wave scattering from small bodies of an arbitrary shape. *Radiofisika 12*, 1185–1197.

Ramm, A. G. (1969b). Iterative solution of the integral equation in potential theory. In *Doklady Academy of Science USSR*, Volume 186, pp. 62–65.

Ramm, A. G. (1970). Approximate formulas for polarizability tensors and capacitances of bodies of arbitrary shapes and applications. In *Doklady Academy of Science USSR*, Volume 195, pp. 1303–1306.

Ramm, A. G. (1971a). Approximate formulas for polarizability tensor and capacitances for bodies of arbitrary shapes. *Radiofisika 14*, 613–620.

Ramm, A. G. (1971b). Calculation of the scattering amplitude for electromagnetic wave scattering by small bodies of arbitrary shapes. *Radiofisika 14*, 1458–1460.

Ramm, A. G. (1971c). Electromagnetic wave scattering by small bodies of an arbitrary shape. *Proceedings of 5th All-union Symposium on Wave Diffraction, Trudy.math.Inst.Steklova, Leningrad*, 176–186.

Ramm, A. G. (1974a). Light scattering matrix for small particles of an arbitrary shape. *Optics and Spectroscopy 37*, 125–129.

Ramm, A. G. (1974b). Scalar scattering by the set of small bodies of an arbitrary shape. *Radiofisika 17*, 1062–1068.

Ramm, A. G. (1980a). Electromagnetic wave scattering by small bodies of arbitrary shapes. In V. Varadan (Ed.), *Acoustic, Electromagnetic and*

Elastic Scattering-Focus on T-matrix Approach, pp. 537–546. Pergamon Press, New York.

Ramm, A. G. (1980b). *Theory and Applications of Some New Classes of Integral Equations.* Springer-Verlag, New York.

Ramm, A. G. (1982). *Iterative Methods for Calculating Static Fields and Wave Scattering by Small Bodies.* Springer-Verlag, New York.

Ramm, A. G. (1986). *Scattering by an Obstacle.* D. Reidel Publishing Company, Dordrecht.

Ramm, A. G. (1987). Completeness of the products of solutions to pde and uniqueness theorems in inverse scattering. *Inverse Problems 3*(4), L77–L82.

Ramm, A. G. (1988a). Multidimensional inverse problems and completeness of the products of solutions to pde. *Journal of Mathematical Analysis and Applications 134*(1), 211–253.

Ramm, A. G. (1988b). Recovery of the potential from fixed-energy scattering data. *Inverse Problems 4*(3), 877.

Ramm, A. G. (1988c). Uniqueness theorems for multidimensional inverse problems with unbounded coefficients. *Journal of Mathematical Analysis and Applications 136*(2), 568–574.

Ramm, A. G. (1989). Multidimensional inverse scattering problems and completeness of the products of solutions to homogeneous pde. *Zeitschrift für angewandte Mathematik und Mechanik 69*(4), T13–T22.

Ramm, A. G. (1992). *Multidimensional Inverse Scattering Problems.* Longman/Wiley, New York.

Ramm, A. G. (1994). *Multidimensional Inverse Scattering Problems.* Mir Publishers, Moscow.

Ramm, A. G. (2000). Finding small inhomogeneities from surface scattering data. *Journal of Inverse and Ill-Posed Problems 8*(2), 205–210.

Ramm, A. G. (2001). A simple proof of the Fredholm alternative and a characterization of the Fredholm operators. *The American Mathematical Monthly 108*(9), 855–860.

Ramm, A. G. (2002). Stability of solutions to inverse scattering problems with fixed-energy data. *Inverse Problems: Mathematical and Analytical Techniques with Applications to Engineering 70*, 97–161.

Ramm, A. G. (2003). A characterization of unbounded Fredholm operators. *Cubo a Mathematical Journal 5*(3), 91–95.

Ramm, A. G. (2005a). *Inverse Problems.* Springer, New York.

Ramm, A. G. (2005b). *Wave Scattering by Small Bodies of Arbitrary Shapes.* World Scientific Publishers, Singapore.

Ramm, A. G. (2007a). Many-body wave scattering by small bodies and applications. *Journal of Mathematical Physics 48*, 103511.

Ramm, A. G. (2007b). Materials with a desired refraction coefficient can be made by embedding small particles. *Physics Letters A 370*(5), 522–527.

Ramm, A. G. (2007c). Scattering by many small bodies and applications to condensed matter physics. *EPL (Europhysics Letters) 80*(4), 44001.

Ramm, A. G. (2007d). Wave scattering by small particles in a medium. *Physics Letters A 367*(1), 156–161.

Ramm, A. G. (2007e). Creating materials with desired properties. In *Oberwolfach Workshop Material Properties, Reports*, Volume 58, pp. 10–13.

Ramm, A. G. (2007f). Distribution of particles which produces a desired radiation pattern. *Communications in Nonlinear Science and Numerical Simulation 12*(7), 1115–1119.

Ramm, A. G. (2007g). Distribution of particles which produces a smart material. *Journal of Statistical Physics 127*(5), 915–934.

Ramm, A. G. (2007h). *Dynamical Systems Method for Solving Operator Equations*. Elsevier, Amsterdam.

Ramm, A. G. (2007i). Electromagnetic wave scattering by many small particles. *Physics Letters A 360*(6), 735–741.

Ramm, A. G. (2007j). Wave scattering by small impedance particles in a medium. *Physics Letters A 368*(1), 164–172.

Ramm, A. G. (2008a). Distribution of particles creating smart material. *International Journal of Tomography & Simulation 8*(8), 25–31.

Ramm, A. G. (2008b). Electromagnetic wave scattering by small bodies. *Physics Letters A 372*(23), 4298–4306.

Ramm, A. G. (2008c). Inverse scattering problem with data at fixed energy and fixed incident direction. *Nonlinear Analysis: Theory, Methods and Applications 69*(4), 1478–1484.

Ramm, A. G. (2008d). A recipe for making materials with negative refraction in acoustics. *Physics Letters A 372*(13), 2319–2321.

Ramm, A. G. (2009a). A collocation method for solving integral equations. *International Journal of Computing Science and Mathematics 2*(3), 222–228.

Ramm, A. G. (2009b). Creating desired potentials by embedding small inhomogeneities. *Journal of Mathematical Physics 50*, 123525.

Ramm, A. G. (2009c). Inverse scattering with non-overdetermined data. *Physics Letters A 373*(33), 2988–2991.

Ramm, A. G. (2009d). Preparing materials with a desired refraction coefficient. *Nonlinear Analysis: Theory, Methods and Applications 71*(12), e186–e190.

Ramm, A. G. (2009e). Preparing materials with a desired refraction coefficient and applications. In C. Skiadas and I. Dimotikalis (Eds.), *Topics in Chaotic Systems: Selected Papers from Chaos 2008 International Conference*, pp. 265–273. World Scientific Publishing Co., Singapore.

Ramm, A. G. (2010a). Electromagnetic wave scattering by many small bodies and creating materials with a desired refraction coefficient. *Progress in Electromagnetics Research M 13*, 203–215.

Ramm, A. G. (2010b). Materials with a desired refraction coefficient can be created by embedding small particles into a given material. *The International Journal of Structural Changes in Solids 2*(2), 17–23.

Ramm, A. G. (2010c). A method for creating materials with a desired refraction coefficient. *International Journal of Modern Physics B 24*(27), 5261–5268.

Ramm, A. G. (2010d). Uniqueness of the solution to inverse scattering problem with backscattering data. *Eurasian Mathematical Journal (EMJ) 1*(3), 97–111.

Ramm, A. G. (2010e). Uniqueness theorem for inverse scattering problem with non-overdetermined data. *Journal of Physics A: Mathematical and Theoretical, Fast Track Communications 43*, 112001.

Ramm, A. G. (2011a). Scattering of electromagnetic waves by many thin cylinders. *Results in Physics 1*(1), 13–16.

Ramm, A. G. (2011b). Scattering of scalar waves by many small particles. *AIP Advances 1*(2), 022135–022135.

Ramm, A. G. (2011c). Uniqueness of the solution to inverse scattering problem with scattering data at a fixed direction of the incident wave. *Journal of Mathematical Physics 52*, 123506.

Ramm, A. G. (2011d). Wave scattering by small bodies and creating materials with a desired refraction coefficient. *Afrika Matematika 22*(1), 33–55.

Ramm, A. G. (2012). A variational principle and its application. *International Journal of Pure and Applied Mathematics 77*(3), 309–313.

Ramm, A. G. (2013a). Electromagnetic wave scattering by small impedance particles of an arbitrary shape. *Journal of Applied Mathematics and Computing (JAMC) 43*(1), 427–444.

Ramm, A. G. (2013b). Heat transfer in a medium in which many small particles are embedded. *Mathematical Modelling of Natural Phenomena 8*(1), 193–199.

Ramm, A. G. (2013c). Many-body wave scattering problems in the case of small scatterers. *Journal of Applied Mathematics and Computing 41*(1–2), 473–500.

Ramm, A. G. (2013d). Scattering of electromagnetic waves by many nanowires. *Mathematics 1*(2), 89–99. Open access Journal: http://www.mdpi.com/journal/mathematics.

Ramm, A. G. (2013e). A variational principle and its application to estimating the electrical capacitance of a perfect conductor. *American Mathematical Monthly 120*(8), 747–751.

Ramm, A. G. (2013f). Wave scattering by many small bodies: Transmission boundary conditions. *Reports on Mathematical Physics* 71(3), 279–290.

Ramm, A. G. (2013g). *Scattering of Acoustic and Electromagnetic Waves by Small Bodies of Arbitrary Shapes — Applications to Creating New Engineered Materials*, Momentum Press, New York, 2013.

Ramm, A. G. (2014a). Inverse scattering with the data at fixed energy and fixed incident direction. *Mathematical Modelling of Natural Phenomena*, 9, N5, 214–253.

Ramm, A. G. and M. Andriychuk (2014b). Application of the asymptotic solution to EM wave scattering problem to creating medium with a prescribed permeability. *Journal of Applied Mathematics and Computing (JAMC)*, 45, 461–485.

Ramm, A. G. (2017). *Creating Materials with a Desired Refraction Coefficient*, IOP Concise Physics, Morgan & Claypool Publishers, San Rafael, CA, USA.

Ramm, A. G. (2018). Many-body wave scattering problems for small scatterers and creating materials with a desired refraction coefficient, in the book *Mathematical Analysis and Applications: Selected Topics*, Wiley, Hoboken NJ, Chapter 3, pp. 57–76 (Ruzhansky, M., H. Dutta, and R. Agarwal (Ed.)).

Ramm, A. G. (2020). *Creating Materials with a Desired Refraction Coefficient*, IOP Publishers, Bristol, UK (Second edition).

Ramm, A. G. (2022). Wave scattering by many small impedance particles and applications, *Reports on Mathematical Physics*, 90, N2 (2022), 193–202.

Ramm, A. G. (2023). Is creating materials with a desired refraction cofficient practically possible? *Characterization and Application of Nanomaterials*, (2023), Vol. 6, N1, 1–5.

Rayleigh, J. W. S. B. (1964). *Scientific Papers*. Dover, New York.

Saito, Y. (1986). An approximation formula in the inverse scattering problem. *Journal of Mathematical Physics* 27, 1145–1153.

Zhikov, V. V., S. M. Kozlov, and O. Oleĭnik (1994). *Homogenization of Differential Operators and Integral Functionals*. Springer Verlag, Berlin.

Index